Springer Protocols Handbooks

For further volumes:
http://www.springer.com/series/8623

Springer Protocols Handbooks collects a diverse range of step-by-step laboratory methods and protocols from across the life and biomedical sciences. Each protocol is provided in the Springer Protocol format: readily-reproducible in a step-by-step fashion. Each protocol opens with an introductory overview, a list of the materials and reagents needed to complete the experiment, and is followed by a detailed procedure supported by a helpful notes section offering tips and tricks of the trade as well as troubleshooting advice. With a focus on large comprehensive protocol collections and an international authorship, Springer Protocols Handbooks are a valuable addition to the laboratory.

Basic Techniques in Biochemistry, Microbiology and Molecular Biology

Principles and Techniques

Aakanchha Jain

Bhagyoday Tirth Pharmacy College, Sagar, Madhya Pradesh, India

Richa Jain

Centre for Scientific Research and Development, People's University, Bhopal, Madhya Pradesh, India

Sourabh Jain

Sagar Institute of Pharmaceutical Sciences, Sagar, Madhya Pradesh, India

Humana Press

Aakanchha Jain
Bhagyoday Tirth Pharmacy College
Sagar, Madhya Pradesh, India

Richa Jain
Centre for Scientific Research and Development
People's University
Bhopal, Madhya Pradesh, India

Sourabh Jain
Sagar Institute of Pharmaceutical Sciences
Sagar, Madhya Pradesh, India

ISSN 1949-2448 ISSN 1949-2456 (electronic)
Springer Protocols Handbooks
ISBN 978-1-4939-9860-9 ISBN 978-1-4939-9861-6 (eBook)
https://doi.org/10.1007/978-1-4939-9861-6

This Humana imprint is published by the registered company Springer Science+Business Media, LLC, part of Springer Nature.
The registered company address is: 233 Spring Street, New York, NY 10013, U.S.A.

DEDICATED TO

MY MENTOR: MY FATHER

"Late Prof. P.C. Jain" and My family

Preface

In today's era of integrated science, the boundaries of different subjects have been dissolved, and the students are merger of all biological sciences. Over centuries, the world of biological sciences has changed radically. Years back, there were no machines like PCR, MALDI-MS, PET, etc. and nearly no plant and animal strains of altered genetic codes, but the present scenario announces new proteins and enzymes every week. With the increasing molecular biology, the basic biochemical, microbiological, and biotechnological analysis needs to be advanced appreciably. Herein, the book *Basic Techniques in Biochemistry, Microbiology and Molecular Biology* will cover experiments in microbiology and molecular biology (for BSc and MSc students), while Section 2 will comprise of the experiments in biochemistry (as per revised new syllabus RGPV and PCI). The present book will provide insight to practicals of biochemistry, microbiology, and molecular biology to build the practical base of subject for readers. Thus, the book will be useful for both undergraduate and postgraduate pharmacy and life science students.

The book will cover applied aspects of practical described herein in addition to the principles of experiment and procedures. The protocols included are brief and clearly defined. Appendices incorporate composition and preparation of media and buffers. The book will also apprise the students with the importance of the subject and acquaint them. The simplified experimental protocols chosen herein will be replicable in college, institute, and university laboratories.

We hope that this will enlighten and instigate the students of all Indian university pharmacy colleges and that the book proposal has vim and vigor to allow its acceptance for publication by Springer Nature.

We welcome suggestions and criticisms from all scientific communities.

Sagar, Madhya Pradesh, India — ***Aakanchha Jain***
Bhopal, Madhya Pradesh, India — ***Richa Jain***
Sagar, Madhya Pradesh, India — ***Sourabh Jain***

Contents

PART IV BIOCHEMISTRY

About the Authors

Aakanchha Jain is an Associate Professor at Bhagyoday Tirth Pharmacy College, Sagar, India. She possesses a throughout first class career. She has to her credit 11 national awards including **Nagarjuna Award 2010, Gour Samman 2010, Ranbaxy Science Scholar 2012, and MP Young Scientist Award 2018.** She is a Zealous Teacher and Researcher. She has published 23 review and research articles in peer-reviewed high-indexed journals. She was formerly INSPIRE DST Research Fellow at the Department of Pharmaceutical Sciences, Dr. H. S. Gour University, Sagar, India, from where she has completed her PhD, MPharm, and BPharm. She has also published two invited book chapters (Wiley, Pan Stanford) covering wide aspects of biotechnology demonstrating her innovation and creativity in delivering lasting results in the field. She has examined many aspects of drug delivery through nanoparticles in her MPharm and PhD. Her main contributions have been in the field of formulation and characterization of novel drug delivery systems, transdermal drug delivery, and anticancer drug delivery systems.

Richa Jain is a Senior Scientist, working in the Centre for Scientific Research and Development, People's University, Bhopal, India, possesses more than 14 years of research experience in the area of molecular biology, animal cell culture, microbiology, bioremediation, and industrial and therapeutic enzymes. Presently, she is working on two projects: (a) therapeutic enzymes arginases (anticancer) and (b) bioremediation of plastic waste. She has published 21 research papers with 81 citations in research journals of high repute. Among several feathers on her cap includes **Best Paper Award** for the presentation of her paper on "Bioinformatics: Importance of Proteomics and Genomics" during the National Conference on "New Frontiers of Sciences" National Science Day Celebrations 2003, Dr. Hari Singh Gour University, Sagar, 28 February 2003; **All India Young Scientist Award** for the presentation of her paper "Production of Proteins and Amino Acids Using Poultry Feathers from a Protease Producing Strain of *Streptomyces sp.* CFS 1068" during Bhartiya Vigyan Sammelan 2007 in Silver Jubilee of All India Young Scientist Conference, M.P. Council of Science and Technology, Bhopal, 23–25 November 2007; **1st Prize** for her paper presentation on "Fluconazole-Loaded Cubosome: An Excellent Nanoparticle for Treatment of Cutaneous Candidiasis" during the National Seminar on Recent Advancement in Drug Delivery Techniques and Drug Designing, School of Pharmacy Research, Peoples University, Bhopal, 26–27 September 2014; and **Appreciation for Oral Presentation** in the International Conference on Translational Medicine in the Twenty-First-Century "Stem Cell Transplantation: Current Status," Barkatullah University, Bhopal, 11–14 April 2015. She has also presented paper entitled "Study of Anti-adenocarcinoma Activity of Arginase by *Streptomyces* Species." She is also a Member of different societies like the Association of Microbiologists of India, Indian Science Congress Association, Institutional Biosafety Committee, etc. She has actively participated in more than 30 scientific workshops and conferences and has also delivered invited guest lectures in workshops and conferences along with providing hands-on training in eight workshops.

Sourabh Jain is an Associate Professor at Sagar Institute of Pharmaceutical Sciences, Sagar, India. He has completed his PhD from Shobhit University, Meerut, UP, India; MPharm from Smriti College of Pharmaceutical Education, Indore, MP, India; and BPharm from B.R. Nahata College of Pharmacy, Mandsaur, MP, India. He is having 8 years of experience in academic and research. He has published 24 papers in national and international journals and conferences. His key research interests include standardization of Ayurvedic and polyherbal formulation, antidiabetic screening and phytochemical isolation of medical plant, and biochemical testing.

Part I

Instruments

Chapter 1

BOD (Biochemical Oxygen Demand or Biological Oxygen Demand) Incubator

Principle: It is also called as refrigerated incubators or low temperature incubators. It is based on the controlled and systemic growth of microorganism at optimum or constant low temperatures.

Application: Biological processing and growth of microbes in aerated conditions

For various microbiological assays, fermentation studies, etc.

Operating Procedure:

1. Connect the electric cord of the unit to voltage stabilizer main socket Photograph 1.
2. Switch ON the main switch.
3. Switch ON the current supply to the unit by turning the ON/OFF switch to ON position.
4. To set the desired working temperature, press the SET push button with finger, and then rotate the coarse fine motion knob until the desired working temperature is indicated on digital controller. The ambient temperature achieved will be displayed ON by L-shape thermometer.
5. Switch ON the air circulating fan to get the uniform temperature in the Chamber.
6. Correct the temperature variation by TP (time proportional) heat-slotted screw. In case the cooling temperature substantially falls below the preset cooling temperature, it will show the green signal, and then rotate this screw in clockwise direction, until the preset temperature is attained. Similarly if the temperature goes beyond the preset temperature, it will show the red signal; then rotate this screw slightly in anticlockwise direction, until the correct working temperature is attained.
7. Switch ON for UV germination tube whenever you required, the switch is given on the controlled board (optional). The BOD is now put into operation.
8. Switch OFF the instrument, when not in use.

Aakanchha Jain et al. *Basic Techniques in Biochemistry, Microbiology and Molecular Biology: Principles and Techniques*, Springer Protocols Handbooks, https://doi.org/10.1007/978-1-4939-9861-6_1,

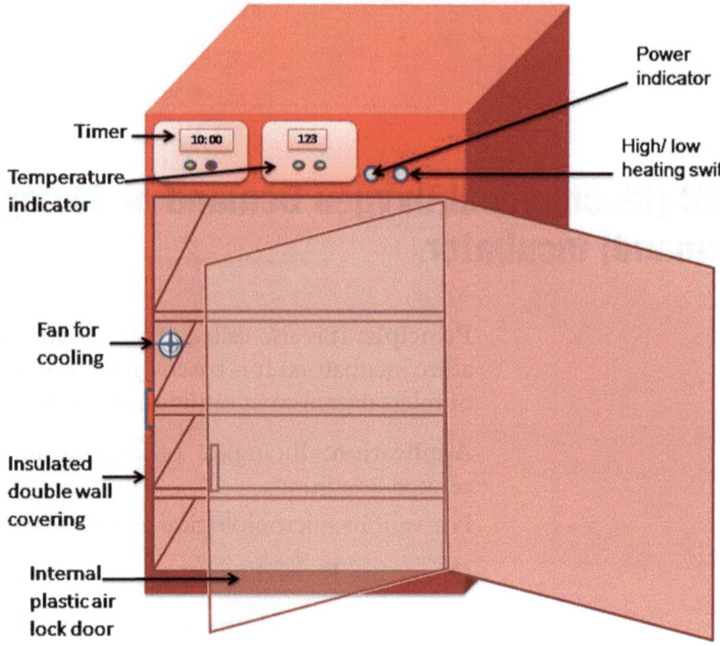

Photograph 1 BOD incubator

Precautions:

1. Incubator must be used with a suitable voltage stabilizer.
2. When the BOD incubator is not intended to use, disconnect it from the current supply.
3. The door of incubator should be opened only when necessary.

Chapter 2

Laminar Air Flow/Biosafety Cabinets

Principle: Laminar flow is mere an indication of the direction of movement of air entering a closed compartment meant for aseptic processing of formulations or at working desk to avoid contamination. The air supplied is via "High-Efficiency Particulate Air" (HEPA) filters. These are filters with pore size of 0.45 μ and having an air velocity of ~99 km/s. Laminar air can be supplied in the desired area horizontally, vertically, or curvilinearly. It has a UV lamp used for sterilization of working desk.

Application: Laminar air flow is used for aseptic transfer of microbial culture.

It is used for providing an excellent aseptic work bench for processes like sub-culturing, sterile drug preparation, inoculations, assays, etc.

Operating Procedure:

1. Install the unit in aseptic room of laboratory.
2. Get the unit cleaned with soft cloth from outside.
3. Fill the manometer with colored water or any other slow-evaporating colored liquid.
4. Connect the electric cord of the unit to main socket, and switch ON the current supply by turning the ON/OFF switch to ON position.
5. Put the UV light ON for about 15 min.
6. Switch ON the blower simultaneously with UV tube.
7. After 15 min switch OFF the UV light, and switch ON the fluorescent tube for lighting.
8. Read pressure drop in the manometer installed on the unit. It should be between 15 and 20 mm. A pressure differential greater than 25 mm indicates clogging of filter.
9. Open the folding door, and clean the table top with cotton soaked in ethyl alcohol.

Aakanchha Jain et al. *Basic Techniques in Biochemistry, Microbiology and Molecular Biology: Principles and Techniques*, Springer Protocols Handbooks, https://doi.org/10.1007/978-1-4939-9861-6_2,

Photograph 1 Laminar Air flow/ Biosafety cabinet

10. Put the Bunsen burner ON.
11. The chamber is now ready for work and continues for the operation.
12. After completion of the work, switch OFF the blower and fluorescent tube light.
13. Close down the folding door (Photograph 1).

Precautions:

1. All material must be sterilized prior to introducing on surface of the laminar air flow workbench.
2. The person should not come in contact with UV light, and efficacy of UV tube light should be checked regularly.
3. Efficiency of HEPA filter should be checked once in a year.

Chapter 3

Aseptic Hood

Principle: The killing of microorganisms by UV light in a closed wooden/steel chamber.

Application: Obsolete method for aseptic culture. Same as laminar air flow.

Operating Procedure:

1. Spray 3.5% acetic acid solution on all surfaces and equipments to meet the criteria for sterility.
2. Sterilize the chamber with UV light 1 h before starting the operation.
3. Fit the gloves on the doors of the cabinet.
4. Place all the glassware inside the cabinet through side door.
5. Now the cabinet is ready for filtration, transfer, and filling of parental product (Fig. 1).

Precautions:

1. Care should be taken to avoid direct contact with UV light.
2. Always wear gloves and mask while handling the parental product.

Aakanchha Jain et al. *Basic Techniques in Biochemistry, Microbiology and Molecular Biology: Principles and Techniques*, Springer Protocols Handbooks, https://doi.org/10.1007/978-1-4939-9861-6_3,

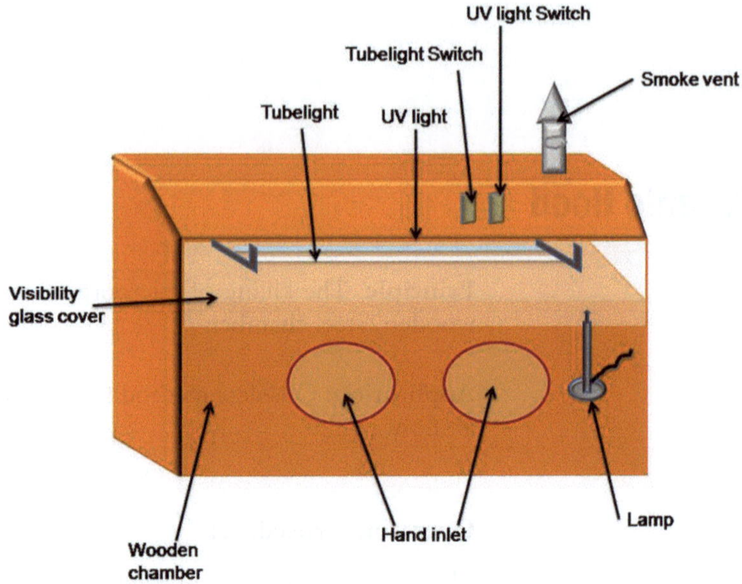

Fig. 1 Aseptic hood

Chapter 4

Autoclave

Principle: It is based on the moist heat sterilization, where pressure at high temperature for definite period of time is used for killing for microorganism in samples which are not destroyed through heat.

Application: Autoclave is used for sterilization of glassware and media for laboratory experiments and for sterilizing instruments used in hospitals, funeral homes, prosthetics, etc.

Operating Procedure:

1. Open the lid by rotating the four-arm steering wheel in anti-clockwise direction, until all the radial arms are disengaged from the grooves.
2. Move the locking arm to the right.
3. Press the foot lifting device, and the lid will be raised up. Turn the lid around, and allow it to rest on the brim of the autoclave.
4. Pour water directly into the boiler drum around 1/4 inches below the bottom of the perforated stand of basket.
5. Replace the lid, and push the locking arm back to the original left position. Now rotate the four-arm steering wheel to tighten the lid in the clockwise direction, ensuring that all the radial arms are securely fixed into the grooves.
6. Connect the electric cord of the unit to the main socket, and switch ON the current supply by turning the ON/OFF switch to ON position.
7. Ensure that the steam release/pressure release valve is securely tight.
8. As soon as the pressure releases a mixture of steam and the air is already present in the boiler reaches 5–7 psi, exhaust it out fully by rotating the steam release valve in anticlockwise direction. After complete exhaustion tighten back the steam release valve.

Aakanchha Jain et al. *Basic Techniques in Biochemistry, Microbiology and Molecular Biology: Principles and Techniques*,
Springer Protocols Handbooks, https://doi.org/10.1007/978-1-4939-9861-6_4,

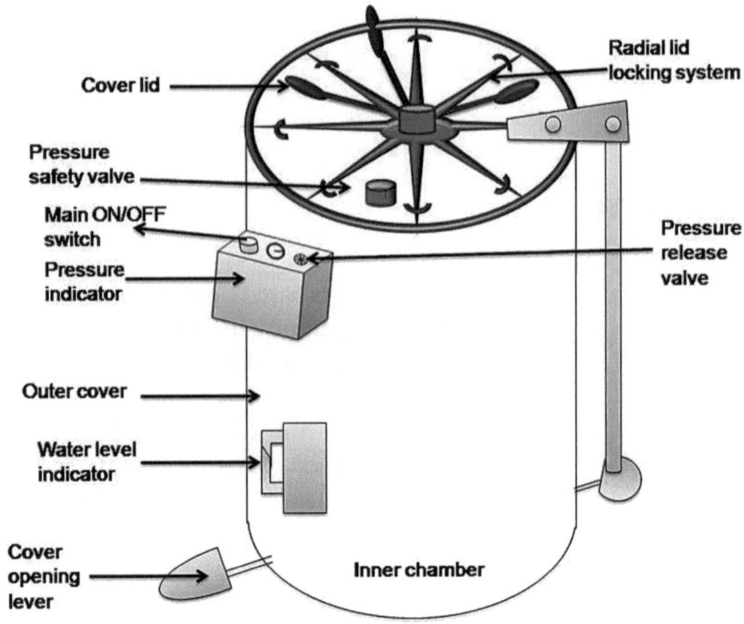

Fig. 1 Autoclave

9. Adjust the pressure by automatic pressure state control switch to 5–20 psi pressure.
10. Keep the autoclave at this pressure for desired length of time to permit proper sterilization of materials.
11. After sterilization has taken place, exhaust out the pressure completely from the boiler drum, move the locking arm to right position, rotate the four-arm steering wheel in anticlockwise direction, and turn around the lid to take out the sterilized materials.
12. Drain out water from the boiler drum (Fig. 1).

Chapter 5

Hot Air Oven

Principle: It involves sterilization by dry heat. The instrument is composed of double-walled chamber of stainless steel or aluminum separated from the outer insulated layer through glass wool filling. Heating is achieved by electric coils filleted inside the double wall.

Application: Hot air oven is used for drying and dry heat sterilization of articles like glassware, forceps, scissors, swab, spatula and pharmaceutical excipients like talc, zinc oxide, and drugs.

Operating Procedure:

1. Switch ON the current supply of oven by keeping the knob on ON position, the red indicator will glow.
2. Keep the preparation/material/article on the perforated self and close the door.
3. Now set the desired temperature by rotating temperature controlling knob; the green indicator will glow.
4. After some time, green indicator goes OFF, indicating that the designed temperature has been reached.
5. Check the temperature from the temperature display unit.
6. After required time period, switch OFF the instrument.
7. Allow to cool to 40 °C before removal of the article.
8. Unplug the unit when not in use (Fig. 1).

Aakanchha Jain et al. *Basic Techniques in Biochemistry, Microbiology and Molecular Biology: Principles and Techniques*, Springer Protocols Handbooks, https://doi.org/10.1007/978-1-4939-9861-6_5,

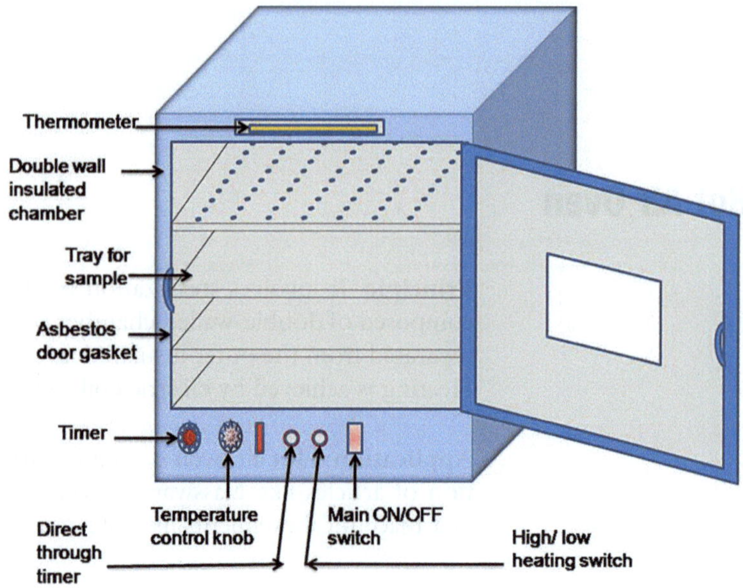

Fig. 1 Hot air sterilizer

Precaution:

1. If temperature does not match with the required temperature, then slightly adjust temperature controller to achieve the desired temperature inside the chamber.
2. Electric current should be disconnected when the unit is not in operation.

Chapter 6

Deep Freezer (−20 °C) (Low-Temperature Cabinet)

Principle: Based on lowering of temperature. The instrument is composed of coolants like liquid nitrogen used for instant and fast freezing of samples.

Application: For quick freezing and pre-freezing of culture, articles, solution and suspension of drugs, and novel drug delivery systems.
For storing of heat labile substances.

Operating Procedure:

1. Connect the electric cord of the unit to the voltage stabilizer.
2. Connect the electric cord of the voltage stabilizer to mains socket.
3. Switch *on* the main switch.
4. Switch *on* the current supply to the unit by turning the *on/off* switch to *on* position.
5. To set the desired working temperature, press the "*set*" push button with finger, and then rotate the coarse fine motion knob until the desired working temperature is indicated on the digital controller.
6. The deep freezer is now put into the operation (Photograph 1).

Precaution:

1. Since frequent high-voltage fluctuations are likely to damage the condensing (refrigeration) unit, the deep freezer must be used with a suitable voltage stabilizer with high/low cutoff system.
2. When the deep freezer is not intended to be used, disconnect it from the current supply.

Aakanchha Jain et al. *Basic Techniques in Biochemistry, Microbiology and Molecular Biology: Principles and Techniques*, Springer Protocols Handbooks, https://doi.org/10.1007/978-1-4939-9861-6_6,

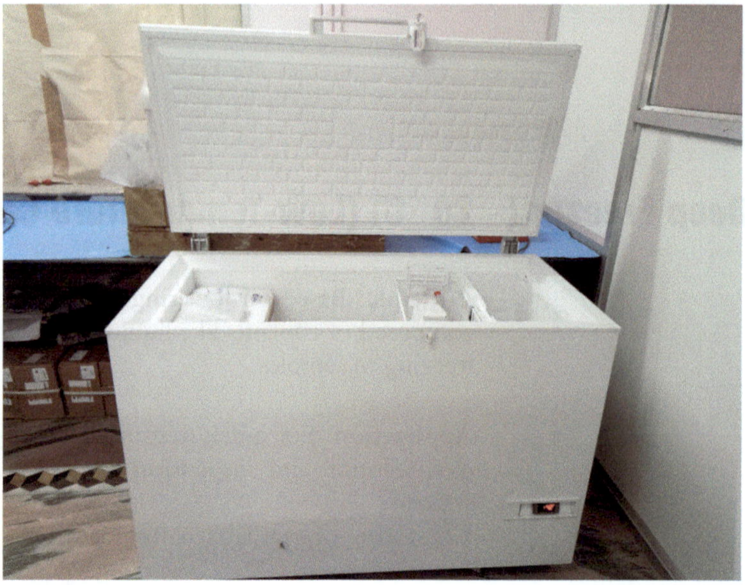

Photograph 1 Deep freezer

3. The refrigerator should be cleaned regularity. When cleaning, turn the power, take out the food in the cabinet, and clean the inside using water or a little neutral detergent.
4. Do not use boiling water, acid, chemical diluents, petrol and volatile oil, or dirt-removing powder.
5. Dry it after cleaning.

Defrosting: Defrost to better freezing efficiency when the frost film in the cabinet 4–5 mm thick.

Chapter 7

Refrigerator

Principle: Based on lowering of temperature. The instrument is composed of coolants like liquid nitrogen used for slow freezing or mere cooling of samples controlled by thermostats.

Application: For preservation of extracts and drug samples which are to be stored at 2–8 °C or till 0 °C.
For storage of immunological products like vaccines, sera, toxoids, toxins, artificial blood and blood products, and novel delivery systems.

Operating Procedure:

1. Connect the electric cord of the unit to the voltage stabilizer.
2. Connect the electric cord of the voltage stabilizer to main socket.
3. Switch ON the main switch.
4. Switch ON the current supply of the unit by turning the ON/OFF switch to ON position.
5. To set the desired working temperature, press the SET push button with finger, and then rotate the coarse fine motion knob until the desired working temperature is set.
6. The refrigerator is now put in to operation.
7. Switch OFF the instrument, when not in use (Fig. 1).

Precautions:

1. Refrigerator must be used with a suitable voltage stabilizer.
2. The refrigerator should be cleaned regularly.
3. Do not keep hot material.

Aakanchha Jain et al. *Basic Techniques in Biochemistry, Microbiology and Molecular Biology: Principles and Techniques*, Springer Protocols Handbooks, https://doi.org/10.1007/978-1-4939-9861-6_7,

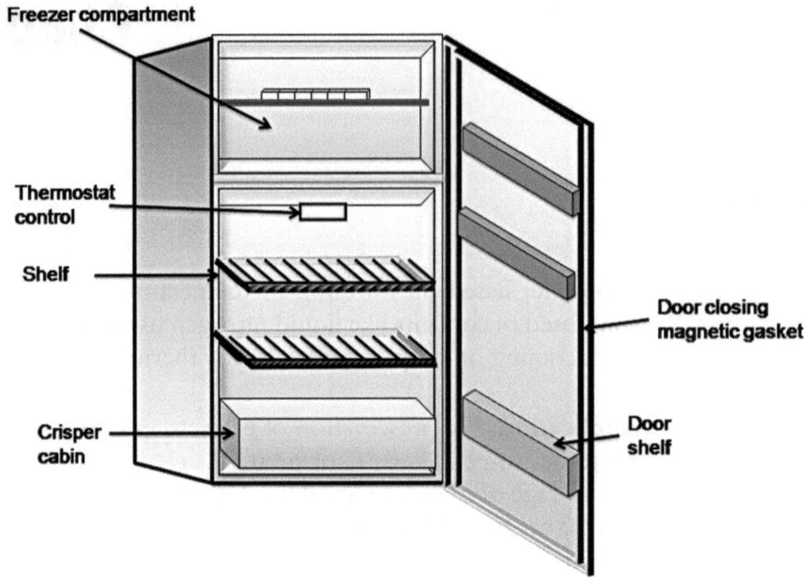

Fig. 1 Refrigerator

Chapter 8

Compound Microscope

Principle: It is an instrument based on magnifying the images using different combination of objective and ocular lens.

Application: For viewing of microorganisms which are not normally visible through naked eyes.
For easy visualization of plant and animal cross (horizontal and vertical) sections.
For anatomical study of microorganism, their internal organelles like mitochondria, flagella, cell wall, etc.

Procedure:

1. Clean the stage and mirror of the instrument.
2. Put the eyepiece of the desired magnification.
3. Out of the three objective lenses, select the desired one.
4. Adjust the position of the mirror so that maximum brightness is observed through the eyepiece.
5. Put your slide over the stage.
6. Rotate the coarse adjustment screw (bigger one) so as to bring the stage to the proper position.
7. Now rotate the fine adjustment screw (smaller one) so as to obtain a fine and clear image of the object.
8. View your slide.
9. Remove the slide after viewing (Photograph 1).

Precautions:

- Always handle the microscope with both the hands.
- Never lift it by its arm.
- Keep the mirror clean.

Aakanchha Jain et al. *Basic Techniques in Biochemistry, Microbiology and Molecular Biology: Principles and Techniques*, Springer Protocols Handbooks, https://doi.org/10.1007/978-1-4939-9861-6_8,

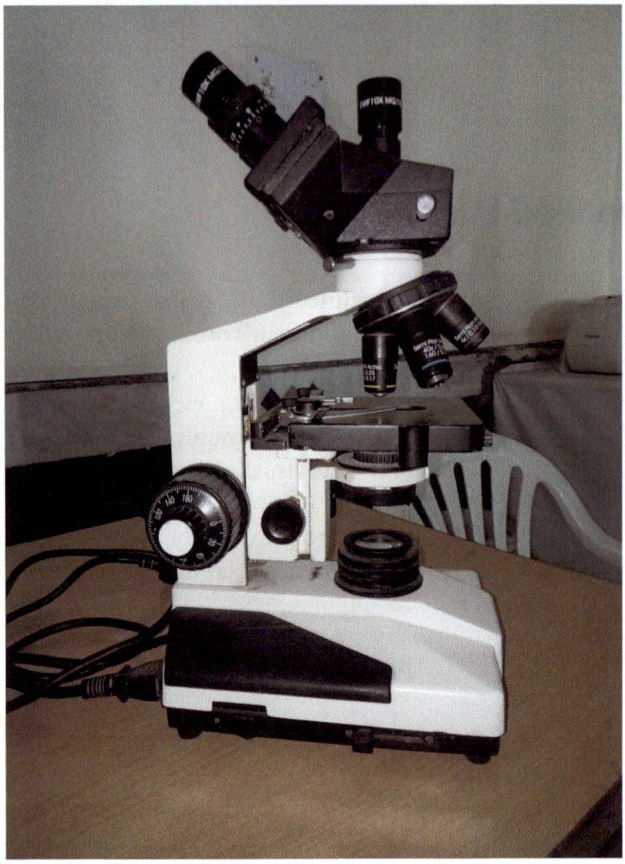

Photograph 1 Compound Microscope

- Adjustment screws should be moved slowly.
- Slide should be very fine and properly stained for good magnification.

Chapter 9

Digital Colony Counter

Principle: It is a microcontrolled instrument which uses a probe/sensor for detection and counting of microbes.

Application: It is used to count the number of colonies on a solid medium in a Petri dish and indirectly for analyzing the total viable and nonviable count of microorganism.

Operating Procedure:

1. Put on the main switch to start on the working of the instrument.
2. Beside this, a tube light inside the instrument grows, to illuminate the counting plate.
3. Put the mounted slide on the counting plate.
4. Adjust the magnifying lens to the suitable position.
5. Now see through the magnifying glass, so that the colony to be counted appears magnified.
6. Count with the help of pen marker.
7. Point each colony with the pen, and press it a little, so that the count appears on the digital display.
8. Count a set of colonies as directed and note down the total count.
9. To count a new set of colony, reset the digital display to zero with the help of reset.
10. Start the counting for new colonies as done previously (Fig. 1).

Precaution:

1. The lens should be fixed at the proper place.

Aakanchha Jain et al. *Basic Techniques in Biochemistry, Microbiology and Molecular Biology: Principles and Techniques*, Springer Protocols Handbooks, https://doi.org/10.1007/978-1-4939-9861-6_9,

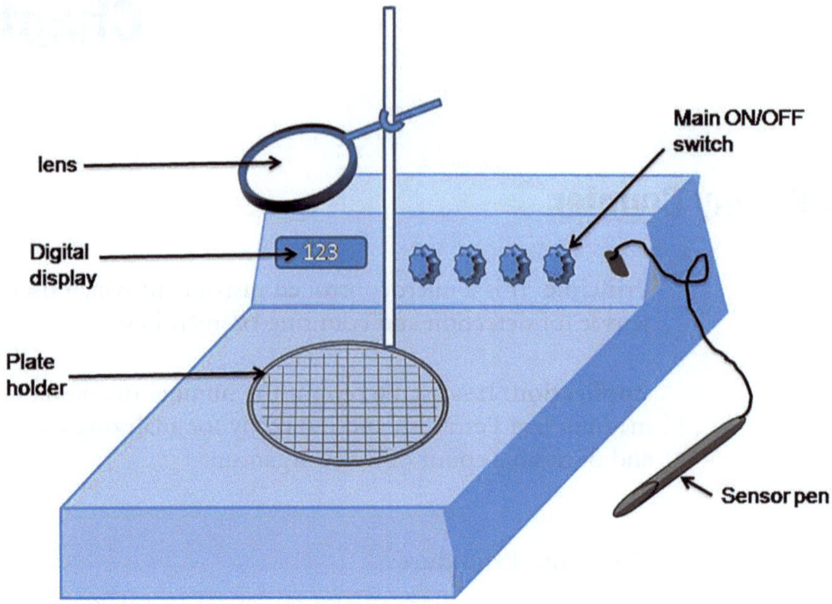

Fig. 1 Colony counter

2. Vertical rod having thread at bottom should be tightened in the bush provided at the top and in the center of colony counter.

Chapter 10

Digital Turbidity Meter

Principle: Turbidity meter is the instrument which measures the decrease in intensity of the transmitted light due to scattering of particles suspended as precipitates or aggregates in a medium. The light passes across the filter creating a light of known wavelength, which is further directed to pass across the unknown sample. Photoelectric cell collects the light and displays it in numerical values on the screen.

Application: It is a method used for determining the concentration of particles suspended as precipitates or aggregates in a medium.
Qualitative limit test for the presence of lead, calcium, etc.
For determining total microbial count and test for antibiotic activity in a sample.

Procedure:

1. Allow sufficient warm-up period after switching on the instrument.
2. Take the test tube containing distilled water or blank solution in the test tube holder, and close the test tube holder lid. Make sure that the mark on test tube coincides with the mark on the panel.
3. Select the required range for measurement.
4. Adjust the display to zero by adjusting "set zero" knob.
5. Remove the test tube containing distilled water, and insert another test tube containing standard solution (say 400 NTU). Place it in the test tube holder.
6. Take the measurement of the solution suspension, and adjust the "calibrate" knob so that the display reads the selected standard solution value.
7. Again check the display zero with the test tube containing distilled water.

Aakanchha Jain et al. *Basic Techniques in Biochemistry, Microbiology and Molecular Biology: Principles and Techniques*, Springer Protocols Handbooks, https://doi.org/10.1007/978-1-4939-9861-6_10,

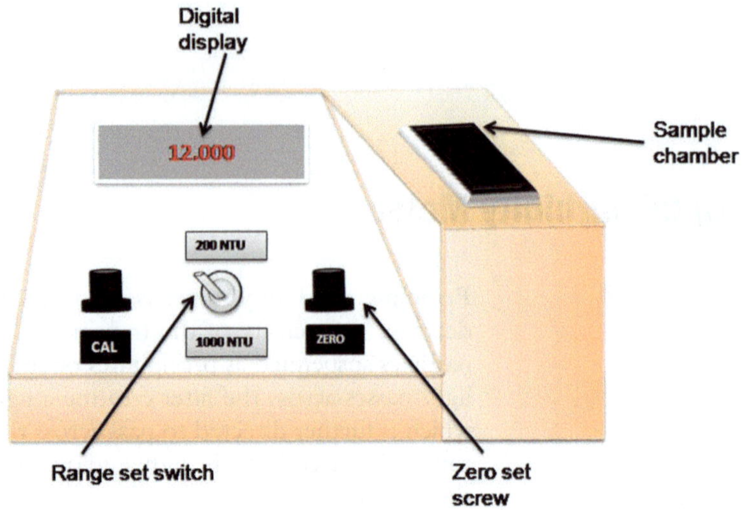

Fig. 1 Turbidometer/Nebular chamber

8. Now the instrument is ready to take measurement of any unknown suspension (Fig. 1).

Precautions:

1. Sample test tube must be thoroughly cleared both inside and outside. In case the test tube gets scratched, discard it.
2. Do not touch the test tube where the light strikes, i.e., at the sides of the test tube, so hold the test tube only at the top end.
3. Fill the test tubes with samples or standard which have been thoroughly to escape otherwise the reading will slowly come down when the reading are taken in the agitated state with air bubbles the reading will be higher.
4. Please ensure that the mark on the test tube coincides with the mark on the instrument panel while taking.

Chapter 11

Digital Nephelometer

Principle and Application: Same as for turbidity meter.

Procedure:

1. Allow sufficient warm up period after switching on the instrument (Fig. 1 in Chap. 10).
2. Take the test tube with distilled water or blank solution. Place it in the test tube holder and close the test tube holder cover. Make sure that the mark on the test tube coincides with the mark on the panel.
3. Adjust the display to 00.0 by adjusting the set zero knob.
4. Now in another test tube, take standard suspension prepared. (Dilute 25 ml of stock turbidity suspension to 100 ml with distilled water. The turbidity of this suspension is 100 NTU.) Prepare this solution weekly.
5. Take its measurement and set the display to the value of the standard suspension with the calibrate knob.
6. Again check the display zero with the test tube containing distilled water.
7. For measurement of turbidity lesser than 20 NTU, thoroughly shake the samples, and wait for unit bubbles to disappear. Put range switch on 20 NTU, and put the sample into the test tube. Take the reading directly from the digital display.
8. For measurement of turbidity above 200 NTU, put the range switch to 200 NTU, and dilute the sample with known volumes of turbidity-free water units; the turbidity falls within 200 NTU. Then compute the turbidity of the original sample from the turbidity of the diluted sample and dilution factor.

Precautions:

1. All the test tubes must be cleaned from both inside and outside before the experiment.

Aakanchha Jain et al. *Basic Techniques in Biochemistry, Microbiology and Molecular Biology: Principles and Techniques*, Springer Protocols Handbooks, https://doi.org/10.1007/978-1-4939-9861-6_11,

2. When tube gets scratched, discard it.
3. Hold the test tube only at the top end. Do not touch the test tube where the light strikes.
4. Fill the test tube with samples or standards, which have been thoroughly agitated.
5. Allow sufficient tine for the air bubbles to escape otherwise the reading will slowly come down. When the reading is taken in the agitated state with air bubbles, the readings will be higher.
6. Make sure that the mark on the test tube coincides with the mark on the instrument panel while taking the readings.

Chapter 12

Digital PhotoColorimeter

Principle: Colorimeter is based on the determination of different gradients of colored compounds in solution. The concentration is measured based on the specific wavelength of light.

Application: It is used for quantitative biochemical tests.

Operating Procedure:

1. Switch on the power supply.
2. Select desired filter by rotating filter disk.
3. Put blank solution or distilled water in the test tube and place it in the holder.
4. Switch on the shutter on switch.
5. Adjust the display to zero optical density by setting O.D. knob.
6. Switch off the shutter switch and remove the given solution from the holder.
7. Measure the reading of O.D.
8. Remove this test tube and place test tube containing solution to be measured.
9. Switch on the shutter switch. Reading says optical density.
10. Concentration of the unknown solution can be calculated as:

$$C = \frac{I_C}{I_S} \times I$$

C = Concentration of unknown solution

I = Optical density of unknown solution

Is = Optical density of standard solution

Ic = Concentration of standard solution

Aakanchha Jain et al. *Basic Techniques in Biochemistry, Microbiology and Molecular Biology: Principles and Techniques*, Springer Protocols Handbooks, https://doi.org/10.1007/978-1-4939-9861-6_12,

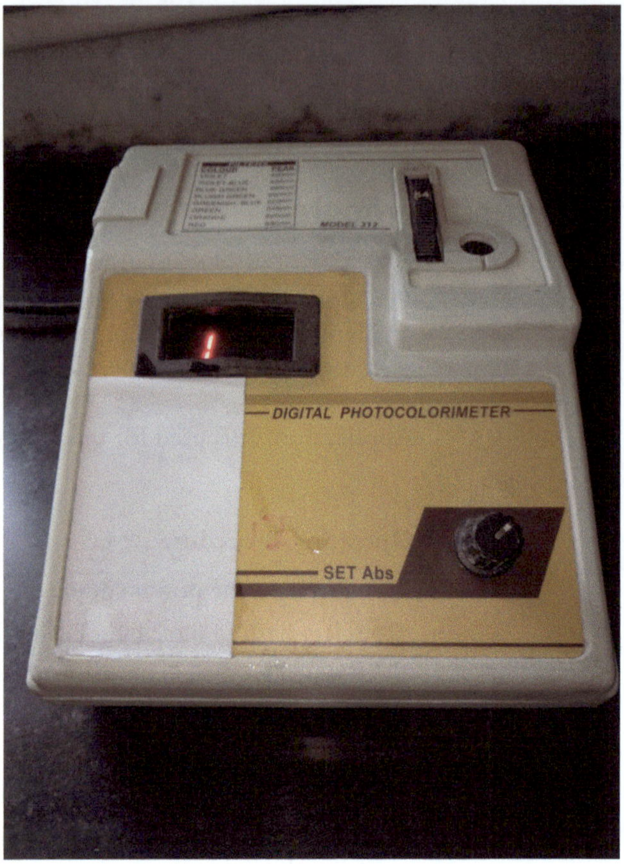

Photograph 1 Colorimeter

Precaution:

1. Display should be zero before the use of the instrument.
2. Allow 5 min warm-up period before starting work on the instrument.
3. Switch off the shutter after use (Photograph 1).

Chapter 13

Digital UV/Visible Spectrophotometer

Principle: The instrument is based on the determination of different concentration of drug/compound using transmitted light in the visible and adjacent ranges. In this region of electromagnetic spectrum, molecules undergo electronic transitions.

Application: It is used for qualitative and quantitative determination of the compound.

Operating Procedure:

1. Switch on the instrument.
2. Display will not appear immediately. First, the monochromatic swing to 190.0 will appear on the wavelength readout.
3. Please wait for 2 min for wavelength display to appear after switching on the instrument.
4. The wavelength can be selected using the two switches on the right side on the instrument, that is, "Set key" and the "Enter key."
5. Now set zero knob to display zero.
6. After setting the desired wavelength, take cuvettes and fill with distilled water, and keep in the cavity in the path of light and zero.
7. After filling any liquid in cuvette, clean with tissue paper and while using different samples. First rinse the cuvette with the sample to test.
8. After finishing the test, switch off the instrument and close the shutter and clean all the cuvette.
9. Switch off the main switch (Fig. 1).

Precaution:

1. Voltage fluctuation should be avoided.
2. Cuvette always cleans with paper.

Aakanchha Jain et al. *Basic Techniques in Biochemistry, Microbiology and Molecular Biology: Principles and Techniques*, Springer Protocols Handbooks, https://doi.org/10.1007/978-1-4939-9861-6_13,

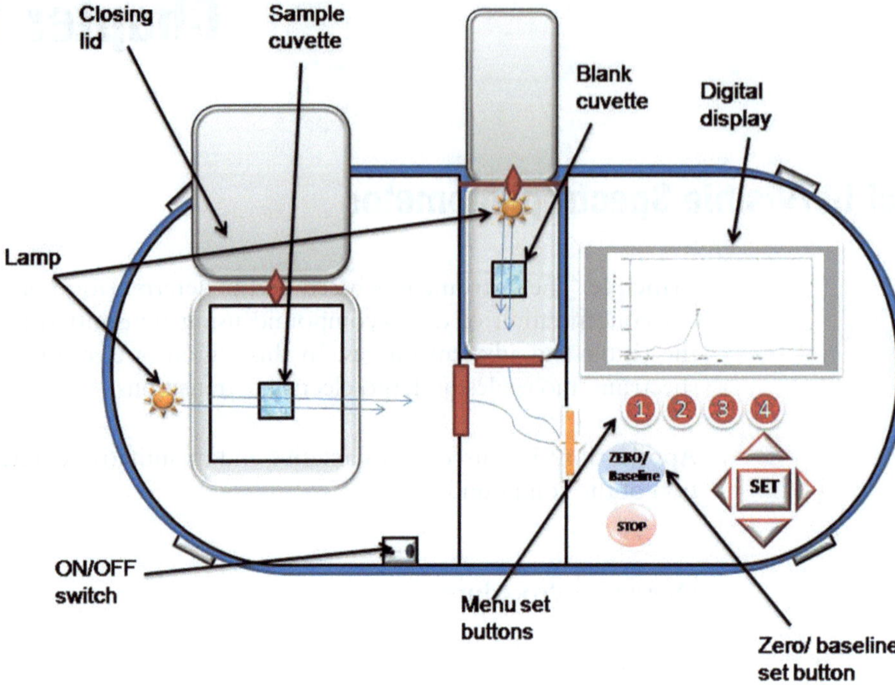

Fig. 1 Spectrophotometer

3. During the experiment mechanical hindrance should be avoided.
4. The environment should be free from dust.
5. The ambient temperature ranging from 15–14 °C.

Chapter 14

Polymerase Chain Reaction

Principle: The instrument is an artificial mode of DNA amplification. Thermal cycler is the machine in which polymerase chain reaction technique is performed. It was invented by Kary Mullis in association with Fred Faloona, Henry A. Erlich, and Randall K. Saiki in the year 1983, while he was working in Emeryville, California, for Cetus Corporation. It is also called as primer-mediated enzymatic amplification of DNA. All the four deoxynucleotides, primer/target DNA with Taq polymerases, and magnesium ions are mitxed and the temperature variated to achieve the three major steps mimicked like body in the instrument: (i) denaturalization, 94–95 °C; (ii) primer annealing, 55–65 °C; and (iii) extension of DNA, 72°. PCR is a technique that takes a specific sequence of DNA of small amounts and amplifies it to be used for further testing.

Applications: In detection of *Neisseria gonorrhoeae*, *Chlamydia trachomatis*, HIV-1, factor V Leiden, fossil identification, forensic testing, and many others.

Operating Procedure: Polypropylene microfuge tubes permit thermal conductivity to wide range of temperature. Thermal cycler is a thermal block having holes; tubes containing PCR mix are kept in it. The cycler alters the temperature of block in pre-programmed patter by reversing the electric current (Peltier effect).

Minimum 1000 copies of DNA template give better results.

Primer is oligonucleotide complementary to 3′ ends of target DNA, Tm (melting temperature) 52–58 °C, GC content: 40–60% (Fig. 1).

Formula for primer Tm calculation is Tm(K) = {ΔH/ ΔS + R ln(C)} or melting temperature Tm(oC) = {ΔH/ΔS + R ln (C)}-273.15, where ΔH (kcal/mole) is change in enthalpy and ΔS (kcal/mole) is change in entropy.

$$\Delta S \text{ (salt correction)} = \Delta S \text{ (1 M NaCl)} + 0.368 \times N \times \ln([Na+])$$

Here, N is the number of nucleotide pairs in the primer (primer length − 1), [Na+] is salt equivalent in mM.

Aakanchha Jain et al. *Basic Techniques in Biochemistry, Microbiology and Molecular Biology: Principles and Techniques*, Springer Protocols Handbooks, https://doi.org/10.1007/978-1-4939-9861-6_14,

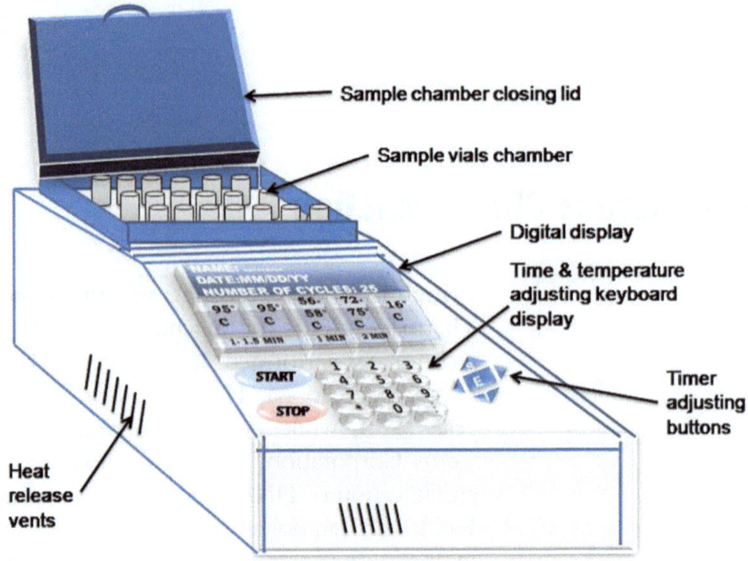

Fig. 1 PCR

The primer annealing temperature is defined by the formula

$$Ta = 0.3 \times Tm\ (\text{primer}) + 0.7\ Tm\ (\text{product}) - 14.9$$

where Tm(primer) = melting temperature of the primers and Tm (product) = melting temperature of the product, 50 mM Tris-HCl (pH 8.3–8.8); $MgCl_2$ (cofactor of DNA polymerase, affects the primer annealing, strand dissociation temperatures of template and PCR product, and enzymatic activity and reliability), KCl to facilitate primer annealing, nuclease-free gelatin or bovine serum acts as stabilizer; distilled water; 1 mM dNTP (deoxynucleotide triphosphates, DNA building blocks). Preferred concentration 20 and 200 uM; DNA polymerase (Taq DNA polymerase, stable at 90 °C, concentration 1–2.5 units (SA = 20 units/pmol) per 100 uL reaction.

For, 25 μL reaction mixture the quantity of reagent required are mentioned below.

10X buffer	2.5 μL
dNTP	0.5
Forward primer	1
Reverse primer	1
Taq polymerase	0.15
Water	18.85 μL
DNA (30–50 ng)	1

Methodology:

1. Switch on the thermocycler.
2. Set the protocol in thermocycler and press the START button.
3. Initialization step (activation of DNA polymerase): raises temperature of the reaction to 94–96 °C or 98 °C for 1–9 min.
4. Denaturation step (breaks hydrogen bonds between bases, releases ssDNA): heating the reaction to 94–98 centigrade for 20–30 s.
5. Annealing step (annealing of primer with ssDNA): 50–65 centigrade for 20–40 s.
6. Extension/elongation step (DNA polymerase adds nucleotide to 3'end of primer): 75–78 centigrade. Generally extension time is 2 min at 72 °C. DNA polymerase will polymerize a thousand bases per minute.
7. Final elongation: 70–74 centigrade for 5–15 min after the last PCR cycle.
8. Final hold: storage of amplified product to 4–15 centigrade.
9. Approximately 35–40 cycles are sufficient for efficient amplification of DNA.
10. Observation: visualize product of PCR by agarose gel electrophoresis.

    Calculation of theoretical output of PCR.

    $Y = X\,(1 + \text{efficiency})^{n}$ Y = amount of amplification target.

    X = input copy number.

    n = number of cycles.

    Efficiency factor is given for each cycle in the kit.

Precautions:

1. Set the protocol in thermal cycler accurately.
2. Mix the reaction mixture under aseptic conditions, and maintain cold chain throughout the process.
3. Take appropriate concentration of template DNA to get best amplification results.

Chapter 15

ELISA Reader

Principle: It is based on the principle of enzyme linked immunesorbant assay where the antigen-antibody reaction is quantitated by means of chromogenic changes on addition of secondary antibody.

Applications: In testing of new immunological products (vaccines) in in vivo samples. In diseases detection, from isolated plasma of patients, in identification of unknown antigen/antibody causing new diseases whose medication is not yet present.

Operating Procedure:

1. Perform ELISA reaction in microtiter (96-well) plate. On completion of procedure, measure optical density using ELISA reader.
2. Press the ON button in the ELISA reader.
3. Switch on the computer attached to ELISA reader,
4. Double click the software program of instrument.
5. Set the wavelength on which the reading has to be measured. Upload protocol.
6. Remove the upper lid/cover of 96 well microtiter plate, and place it firmly onto carriage.
7. Click on the START command displayed in computer.
8. Automatically, plate will be carried inside the instrument, and plate is agitated to mix content of well.
9. Strip is placed exactly below the optical path. (Lamp produces white light that is focused as beam by lens; it passes through the well containing sample. Sample absorbs certain wavelength of light, while the unabsorbed is transmitted. The transmitted light is filtered and focused onto photodiodes that convert received light into electrical signal (digital form) that is transformed into reading (absorbance).)

Aakanchha Jain et al. *Basic Techniques in Biochemistry, Microbiology and Molecular Biology: Principles and Techniques*, Springer Protocols Handbooks, https://doi.org/10.1007/978-1-4939-9861-6_15,

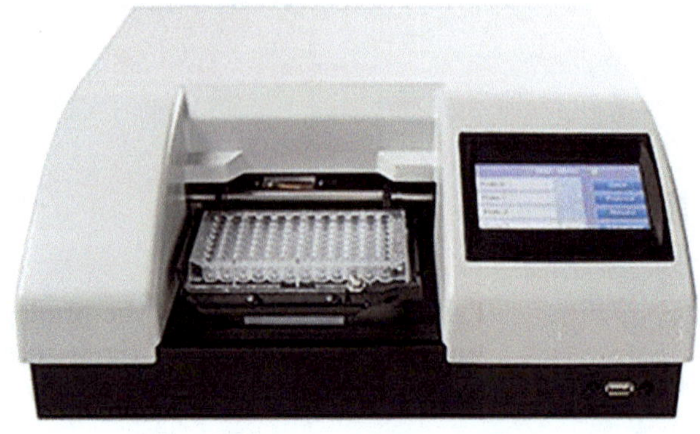

Photograph 1 ELISA reader

10. Measurement is carried out in sequential way from one well to another well.
11. Once the readings are completed, results in terms of optical density/absorbance are displayed on computer screen and printed by printer attached with ELISA reader.
12. Click the STOP button, and the plate will be ejected out; remove plate from carriage.
13. Exit the program.
14. Switch OFF ELISA reader and then computer (Photograph 1).

Precaution:

- Voltage stabilizers should be used to prevent fluctuation in voltage.
- Always handle plate from side, avoid touching it from bottom. Clean the plate bottom with tissue paper before placing it on carriage.
- Switch ON the machine first and then start the program.
- Ensure no leakage of sample from plates while handling.
- The instrument should be kept in cool and dust-free environment.

Chapter 16

Sonicator

Principle: The instrument is based on the use of high-frequency ultrasonic waves which can travel across cell and membranes. It can be of two types: (i) probe and (ii) bath.

Application: For plant cell organelles isolation; reducing particle size of novel drug delivery carriers like liposomes, ethosomes, tansferosomes, and lipid nanoparticle; preparation and formation of novel drug delivery carriers; and killing of microbes in heat-labile preparations.

Operating Procedure:

1. Turn on the power.
2. To change the sonicator parameters, press the PROG button. To change a parameter, type in the new setting, and then press the YES/ENTER button.

   The first time parameter is the amount of time per sample. If you will be sonicating individual samples, this time does not matter, as long as it is longer than the total pulse time.

   The second time parameter (ON) is the length of each pulse.

   The third time parameter (OFF) is the amount of time between pulses.

   The output setting goes from 1 to 10, with 1 being the least power per pulse and 10 being the most. For small (<1 ml) bacterial samples, an output of 3–4 is usually enough.

   The temperature is not controlled; press NO for this one.
3. Put the sonicator tip in your sample tube and press the START/STOP button.
4. After the desired number of pulses, press the START/STOP button again.

Aakanchha Jain et al. *Basic Techniques in Biochemistry, Microbiology and Molecular Biology: Principles and Techniques*, Springer Protocols Handbooks, https://doi.org/10.1007/978-1-4939-9861-6_16,

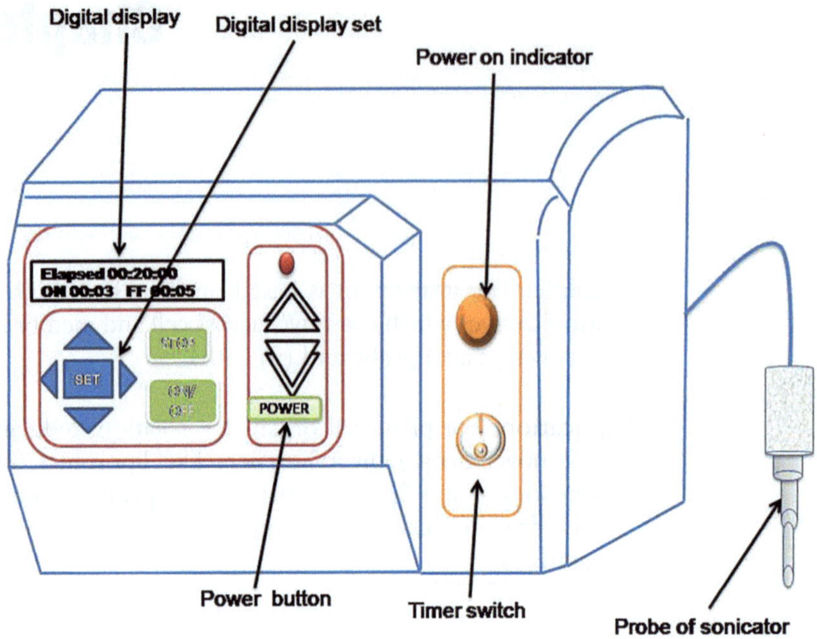

Fig. 1 Probe Sonicator

∗∗Clean the sonicator tip with wipes and ethanol before and after use∗∗

The power button is at the back of the machine on the right side.

Samples should be kept on ice and, if they are pigments, in dim light whenever possible (Fig. 1).

Part II

Molecular Biology

Chapter 17

Enzyme Assay: Qualitative and Quantitative

Enzymes are protein catalyst found in all living forms. Enzyme can be purely a protein and/or a combination of protein (apoenzyme) and nonprotein component (cofactor) essential for catalytic activity. Apoenzyme when attached to its cofactor is defined as holoenzyme. They possess high specificity for the reaction catalyzed and its substrate. Like catalyst they lower down the activation energy of biochemical reaction so that the product formation becomes feasible under normal physiological conditions. They increase the rate of biochemical reaction without being consumed. Biochemically they are large globular proteins with active site where the substrate binds and is converted into product. Each enzyme demonstrates its maximum activity at its respective optimum conditions such as temperature, pH, nature of active site, strength of ions, etc. Even slight divergence from the optimum condition may result in alteration in enzymatic reactions thereby affecting the product formation. The comparison of enzyme activity and replicable results can be obtained only if these conditions are exactingly regarded.

Most of the commercially available, industrially useful and therapeutic enzymes are derived from microorganisms. The reason being they replace the harmful chemicals thereby reducing the environmental pollution (Table 1).

1 Qualitative Enzyme Assays

Enzymes can be easily recognized by their catalytic reactions. Hence, enzyme assays are based on measurement of product formed or disappearance of substrate. Qualitative and quantitative approach can be opted for detection of specific enzyme through

Aakanchha Jain et al. *Basic Techniques in Biochemistry, Microbiology and Molecular Biology: Principles and Techniques*, Springer Protocols Handbooks, https://doi.org/10.1007/978-1-4939-9861-6_17,

Table 1
Classification of enzymes

Type of enzyme	Reaction	Example	Enzyme Commission no. (EC)
Oxidoreductase	Redox reaction	Peroxidases, dehydrogenase	EC 1.1–1.21; 1.97
Transferase	Transfer of groups, such acetyl, amino, etc., between molecules	Aspartate carbamoyl transferase, aminotransferase	EC 2.1–2.10
Hydrolase	Addition or removal of water molecule	Protease, lipase, nuclease	EC 3.1–EC 3.13
Lyase	Addition or removal of group without involvement of water molecule	Threonine aldolase, phosphoenolpyruvate carboxylase	EC 4.1–EC 4.99
Isomerase	Converts molecule in its isomeric form	Methionine racemase, tartrate epimerase	EC 5.1–5.5
Ligase	Energy from ATP is used for joining two molecules	DNA ligase, arginine-succinate synthetase	EC 6.1–EC 6.6

positive or negative reactions. In qualitative assays range of spectra of utility of microorganisms can be depicted, while quantitative method relates to product produce. Detection of product formed is one of the most accepted methods of quantifying enzymes. In condition where more than two products are formed, quantification of any of the product is considered as appropriate for determining enzyme activity. However, more accurate determination may require quantification of both the products. Exact calculation of quantity of enzyme from the product is a complex procedure. Accumulation of product may occur even if the quantity of enzyme may be less hence calculation of specific activity is more accurate measurement of enzyme activity.

Precautions to Be Followed for all Enzymatic Assays:

- Dilutions of enzyme solution to be prepared fresh, before use.
- The enzyme samples should be kept in ice until used.
- The dilutions of enzymes should be prepared in assay buffer only.
- Clean glass tube, microtips, and pipettes without any detergent residue should be used.

- Every experiment should be done in triplicate at a time.
- Assay media plates should be poured and kept in such a way that the width of media should be same all over the plate.
- The fresh microtips or glass pipettes should be used for pipetting.
- Prepare protocol flow chart before performing any experiment.
- Labeling should be done using permanent glass marker.

Protease Assay: The basic principle underlying the measurement of proteolytic activity is based on measurement of products of protein hydrolysis such as peptides and amino acids. Qualitative assay used for assay of proteolytic activity is agar plate assay supplemented with specific protein such as casein, milk, gelatine, and azo dye-impregnated proteins. These assay measures protease activity in terms of zone of clearance (protein hydrolysis) formed around the colony in assay plates.

Amylase Assay: Starch is polysaccharide containing more than eight monosaccharide units. Starch-hydrolyzing enzymes are called as amylases. Detection of amylase is based on the principle that starch in presence of iodine forms starch-iodine complex (blue color) in starch agar medium. In presence of amylase enzyme, starch present in medium gets hydrolyzed into monosaccharide; thereby a colorless zone of enzyme activity is formed around the colony of microorganism.

Material required: Starch agar medium, Gram's iodine solution, sterilized Petri plates, dropper, Bunsen burner, inoculation loop, cork borer, laminar airflow, distilled water, spirit.

Sterilization: Medium: All the medium components are weighed in proportion as described in Table 2. After proper mixing of all media components, check and set the pH as desired using 0.1 N HCl or 0.1 N NaOH. Thereafter sterilize at 121 °C, 15 lbs. for 15 min in autoclave. After sterilization process the medium is allowed to cool under room temperature, when temperature reaches 45 °C (bearable to keep on palm), and then pour 25 ml of medium in each Petri plate. Leave the plates covered in laminar airflow for 24 h under aseptic conditions for inoculation of microorganisms.

Glasswares: All the glasswares like Petri plates, pipette, and beakers are wrapped in paper and kept in hot air oven set at 100 °C for 5–6 h.

Table 2
Medium composition for assay of various enzymes

S.no.	Enzyme	Composition of medium	Sub-component	Detecting reagent
1.	Protease	Casein agar medium: Part A: dissolve 20.0 g skimmed milk in 200 ml distilled water Part B: 20.0 g agar-agar dissolved in 800 ml 0.2 M phosphate buffer (pH 7.0) Mix part A and B after sterilization (mix at 45 °C)	Phosphate buffer (pH 7): mix 39.0 monobasic sodium phosphate solution (0.2 M) and 61 ml of dibasic sodium phosphate solution (0.2 M)	20% (w/v) trichloroacetic acid solution
2.	Amylase	Starch agar medium Soluble starch 20 g, peptone 4.5 g, beef extract 3.0 g, agar-agar 20 g, distilled water 1000 ml	Gram's iodine Iodine 1.0 g, KI 1.5 g, distilled water 300 ml Colorless to yellow zone of enzyme activity is formed against the dark blue background	Gram's iodine
3.	Pectinolytic enzyme	Part A: yeast extract 1.2 g, agar-agar 20 g, pectin 5 g, distilled water 500 ml Part B: mineral salt solution $(NH_4)SO_4$ 2.0 g, KH_2PO_4 2.0 g, Na_2HPO_4 6 g, $FeSO_4.7H_2O$ 0.2 g, $CaCl_2$ 1 mg, H_3BO_3 10 μg, $MnSO_4$ 10 μg, Zn SO_4 70 μg, MoO_3 10 μg, distilled water 500 ml, pH 7	Colorless zone of enzyme activity is formed against the dark background	Detecting reagent: 1% aqueous solution of hexadecyltrimethylammonium bromide
4.	Chitinase	Chitinase assay medium: colloidal chitin 4.5 g, $MgSO_4.7H_2O$ 0.25 g, $(NH_4)SO_4$ 2.0 g, KH_2PO_4 2.0 g, citric acid	Preparation of colloidal chitin: dissolve 4.0 g of commercially available chitin in 40 ml conc HCl. Mix well at 4 °C for	Bromocresol purple dye changes yellow color of the media (in acidic here pH 4.7) into purple color due to increase in pH

(continued)

Table 2 (continued)

S.no.	Enzyme	Composition of medium	Sub-component	Detecting reagent
		monohydrate 1.0 g, agar-agar 20 g, distilled water 1000 ml; bromocresol purple 0.10 g, and Tween 80 2.0 ml; adjust pH 4.7 with the I N HCl	8–10 hours. Neutralize resultant slurry with 100 ml deionized water. Keep slurry in vacuum concentrator to evaporate the water. Check pH of slurry and again repeat the process until pH becomes 3.5. Use this colloidal chitin for media preparation	
5.	Keratinase	Part A: K_2HPO_4 1.5 g, $MgSO_4.7H_2O$ 0.25 g, $CaCl_2.2H_2O$ 0.25 g, $FeSO_4.7H_2O$ 0.015 g, $ZnSO_4.7H_2O$ 0.005 g, agar-agar 14 g, distilled water 1000 ml (pH = 7) Part B: keratin azure 10 mg per 10 ml medium	Prepare basal media stabs as shown, and overlay it with sterilized skeratin azure medium (part B)	Color of medium changes to blue due to release of azo dye

Agar Diffusion Technique:

1. Inoculate (line streak) log phase (actively growing) culture of bacteria/actinomycetes on the equatorial positions of plate (Fig. 1a). Fungi are point inoculated at the center of plate (Fig. 1c).
2. Incubate culture of bacteria at 37 ± 2 °C for 24 to 48 h and those of fungi and actinomycetes at 28 ± 2 °C for 48–72 h in incubators.
3. The growth diameter (width) and zone of enzyme activity diameter will be measured in mm using calibrated scale (Fig. 1b, d).

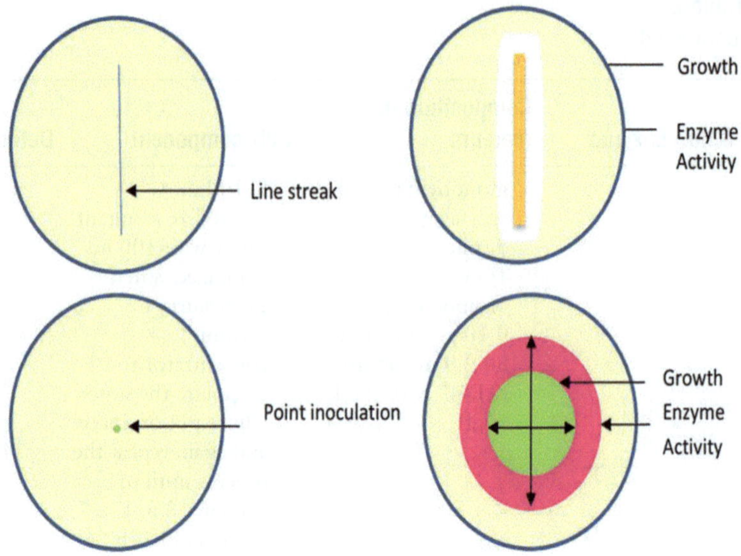

Fig. 1 Agar diffusion assay (**a**) Line Streak of microorganism (Bacteria and Actinomycetes) at the centre of plate; (**b**) Zone of enzyme activity around the growth of microorganism after 24–48 h of incubation at 37±2 °C (Bacteria) and 28±2 °C (Actinomycetes); (**c**) Point inoculation of fungi at the centre of plate; (**d**) Zone of enzyme activity around fungal growth after 72 h of incubation at 28±2 °C

Calculation:

$$\text{Relative Enzyme Activity (REA)} = \frac{\text{Diameter of zone of enzyme activity in mm}}{\text{Diameter of growth in mm}} \quad (1)$$

2 Quantitative Enzyme Assay

Penicillinase assay by macroiodometric method (Sargent 1968)

Materials: Stock solution of penicillin: 1.2 mg per ml (filter sterilize using 22 μ membrane filter), 0.5 M acetate buffer (4.0 pH), iodine stock solution (mix 0.32 N iodine and 1.2 M potassium iodine in 100 ml distilled water), iodine working reagent (5% of stock solution in acetate buffer pH 4.0).

Method:

1. Reaction mixture: To 1 ml of test enzyme sample, add 2.5 ml of phosphate buffer (0.1 M, pH 7.0), and incubate for 10 min at 28 °C. Add 1 ml of penicillin as substrate followed by incubation for 5 min. Terminate the reaction by adding 5 ml of iodine reagent.

2. Blanks: 2.5 ml of phosphate buffer, 1 ml penicillin (substrate), 5 ml of iodine, and 1 ml of test enzyme sample are mixed in sequence mentioned herein before the incubation.
3. Measure change in optical density at 490 nm using spectrophotometer. "1 Unit of Beta-Lactamase activity is defined as quantity of enzyme that hydrolysed 1μ mol of substrate min-1."

   Beta-lactamase activity (units ml-1 min-1).

$$\Delta \text{Ku} \times 40 \times 1 \times 1 \text{ Ku} - \text{blank A F T V}$$

In this equation, ΔKu = (Ku-blank B) minus (Ku- sample), F = consumed moles of iodine per mole of hydrolyzed substrate, T = time (mins) for enzyme reaction, and V = volume of enzyme solution added to the assay tube.

Protease Assay: The casein being one of the simple proteins is used as substrate for protease assay.

Materials:

1. Buffered casein solution 1% (w/v), 0.2 M phosphate buffer pH 7.0, trichloroacetic acid (TCA) solution (20%).
2. Reaction tube: Pipette 1 ml of buffered casein in clean glass test tube.
3. Add 1 ml of enzyme sample to it and mix thoroughly the content using vortex mixer.
4. Incubate the tubes on water bath for 1 h at desired temperature (generally used temperature is 28 ± 2 °C).
5. Stop the reaction by addition of 1 ml of TCA (20%) solution. Mix thoroughly using vortex mixer for 1 min.
6. Keep reaction tubes at 4 °C for 10 min, and then transfer the entire reaction mixture to Eppendorf tube, and centrifuge at 8000 rpm for 15 min at 4 °C.
7. Blank tube: Pipette 1 ml of casein, 1 ml of TCA (20%), and 1 ml of enzyme sample in sequence mentioned herein, followed by incubation at desired temperature. The content of blank tube is then centrifuged as mentioned in reaction tube (Step 6).
8. The tyrosine liberated due to enzyme action is measured using method of Lowry et al. (1951).
9. 1 unit of enzyme activity: Amount of enzyme which liberates 1 μg of tyrosine from casein in 1 hour at 28 °C.
10. (The concentration of tyrosine released in test samples was calculated with the help of standard curve of tyrosine.)

Straight line equation: $y = mx + c$ where y = optical density, x = concentration of tyrosine, and c = slope.

$$\text{Enzyme Units} = \text{T-C} \times \frac{1}{t} \times \frac{V_m}{V_e} \quad (2)$$

T-C = Amount of tyrosine liberated

t = Time in hour

V_m = Volume of reaction mixture

V_e = Volume of enzyme

Standard Curve of Tyrosine Materials:

Solution A: 2% Na_2CO_3 in 0.1 N NaOH.

Solution B: 0.5% $CuSO_4.5H_2O$ in 1% solution of sodium potassium tartrate.

Solution C: mixing 50 ml of *solution A* and 1 ml of *solution B.*

Folin-Ciocalteu phenol reagent: Dilute the commercial reagent with an equal volume of water (1:1) on the day of use.

Tyrosine stock solution (1000 μg per ml).

Procedure:

1. Prepare different concentrations of tyrosine by diluting the stock solution with distilled water as mentioned in Table 3.
2. To 1 ml of different dilutions of tyrosine, add 5 ml of alkaline solution and vortex for 1 min, and then incubate at room temperature for 10 min.
3. Add 0.2 ml of diluted (1:1) Folin's reagent and mix thoroughly for a minute. Cover mouth of tubes with aluminum foil.
4. Incubate the tubes in dark for 30 min at room temperature.
5. Measure absorbance (optical density (OD)) at 660 nm using colorimeter or spectrophotometer, and note readings in Table 3.
6. Plot graph by taking concentrations on x-axis and OD on y-axis. Calculate regression of curve.
7. Note regression value of 0.99 or more is accepted for standard curve. The regression value less than 0.99 indicates pipetting error; thus the experiment is to be repeated.

Amylase Activity Materials: Clean glass test tubes, buffered soluble starch 1% (0.1 M phosphate buffer pH 7.0); 3,5 dinitrosalicylic acid (DNSA) reagent (DNSA 1 g, phenol 0.2 g, sodium sulfite 0.05 g, sodium hydroxide 1 g; dilute to 100 ml using distilled water).

Procedure:

1. Blank tube: Pipette 0.5 ml of soluble starch, 1 ml of DNSA reagent, and 0.5 ml of enzyme sample. Thereafter incubate the tube at desired temperature for 15 min in water bath.
2. Reaction tube: Pipette 0.5 ml of soluble starch and 0.5 ml of enzyme sample. Incubate the tubes in water bath at desired

Table 3
Standard curve of tyrosine

Concentration of tyrosine (μg/ml)	Amount of tyrosine stock solution (μl)	Amount of distilled water	Amount of alkaline solution in ml	Amount of Folin's reagent (1:1 dilution) in ml	Optical density at 660 nm
50	50	950	5	0.2	
100	100	900	5	0.2	
200	200	800	5	0.2	
300	300	700	5	0.2	
400	400	600	5	0.2	
500	500	500	5	0.2	
600	600	400	5	0.2	
700	700	300	5	0.2	
800	800	200	5	0.2	
900	900	100	5	0.2	
1000	1000	0	5	0.2	
0 (blank)	0	1000	5	0.2	

temperature for 15 min. Stop the reaction by adding 1 ml of DNSA reagent.

3. Boil blank tube and reaction tube in water bath for 10–15 min.
4. Cool the tubes at room temperature, and then add 10 ml of distilled water to it with vortex mixing for 1 min.
5. The reducing sugar liberated reacts with DNSA and reduces it; thereby color changes from yellow to orange. The amount of reducing sugar released in reaction mixture is read at 540 nm.
6. One unit (U) of amylase activity is defined as amount of maltose equivalents liberated per min per ml of enzyme under assay conditions.
7. Prepare standard curve of maltose to quantify amount of maltose present in reaction and blank tubes (Table 4).

Chitinase Assay: Chitin is second most abundant natural polymer, composed of repeated units of N-acetylglucosamine, linked with β-1,4-glycosidic bonds. It is generally found in insects' exoskeleton, fungal cell walls, crustacean shells, etc. Chitinases hydrolyze chitin and find application in pharmaceutical industries, agricultural sector, and clinical biology. They are also employed for synthesis of chitin derivatives. Thus novel-specific chitinase with better catalytic activity is in demand.

Table 4
Standard curve of maltose

Concentration of maltose (μg/ml)	Amount of maltose stock solution (μl)	Amount of distilled water	Amount of DNSA reagent (ml)	Optical density at 540 nm
50	50	950	1	
100	100	900	1	
200	200	800	1	
300	300	700	1	
400	400	600	1	
500	500	500	1	
600	600	400	1	
700	700	300	1	
800	800	200	1	
900	900	100	1	
1000	1000	0	1	
0 (blank)	0	1000	1	

Calculate the value using Eq. 2 except T-C = amount of maltose liberated

Materials: Colloidal chitin, DNSA reagent (DNSA 1 g, phenol 0.2 g, sodium sulfite 0.05 g, sodium hydroxide 1 g; dilute to 100 ml using distilled water), phosphate buffer pH 7.0, enzyme sample, reducing sugar (glucose, etc.).

Procedure:

1. Blank tube: Pipette 0.5 ml of colloidal chitin, 1 ml of DNSA reagent, and 0.5 ml of enzyme sample; mix thoroughly for 1 min.
2. Incubate the tube at desired temperature for 15 min in water bath.
3. Reaction tube: Pipette 0.5 ml of colloidal chitin and 0.5 ml of enzyme sample. Incubate the tubes in water bath at desired temperature for 15 min. Stop the reaction by adding 1 ml of DNSA reagent.
4. All the tubes are boiled at 100 °C in blank tube and reaction tube in water bath for 10–15 min.
5. Cool the tubes at room temperature, then add 10 ml of distilled water to it, and mix well.

Table 5
Observation table for standard curve of glucose

Concentration of reducing sugar to be prepared (μmol/ml)	Amount of reducing sugar stock solution (μl)	Amount of distilled water	Amount of DNSA reagent (ml)	Optical density at 540 nm
50	50	950	1	
100	100	900	1	
200	200	800	1	
300	300	700	1	
400	400	600	1	
500	500	500	1	
600	600	400	1	
700	700	300	1	
800	800	200	1	
900	900	100	1	
1000	1000	0	1	
0 (blank)	0	1000	1	

Calculate the value using Eq. 2 except T-C = Amount of reducing sugar liberated

6. The reducing sugar liberated reacts with DNSA and reduces it; thereby color changes from yellow to orange. The amount of reducing sugar released in reaction mixture is read at 540 nm.
7. One unit of chitinase activity was defined as the amount of enzyme that produced 1 μmol of reducing sugars per minute.
8. One unit (U) of activity is defined as amount of maltose equivalents liberated per min per ml of enzyme under assay conditions.
9. Prepare standard curve of glucose to quantify amount of glucose present in reaction and blank tubes (Table 5).

Pectinase Assay Materials: 0.1 M citrate buffer, reagent A (4.0 g of sodium carbonate is dissolved in 60 ml of distilled water. After complete dissolution add 1.6 g of glycine and 0.04 g copper sulfate to it, mix well using magnetic stirrer, and make the volume to 100 ml deionized water), reagent B (0.12 g neocuproine-HCl, 100 ml deionized water), D-galacturonic acid stock 1 mg/ml, substrate (0.5% polygalacturonic acid in water, dissolve by heating).

Procedure:

Reaction tube: To 6 ml substrate (polygalacturonic acid), add 1 ml of enzyme sample.

Enzyme blank: Mix 6 ml substrate (polygalacturonic acid), 1 ml citrate buffer, and 1 ml enzyme sample.

Incubate tubes at 37 °C for 60 min.

Prepare standard curve of D-galacturonic acid in range of 10–100 μg/ml concentration.

Transfer tubes at 4 °C after completion of incubation period, to stop the reaction.

To tubes add 2 ml of reagent A and 2 ml of reagent B, and mix well.

Boil all the tubes in water bath for 15 min.

Cool the tubes at room temperature, and then add 2 ml of distilled water in each tube.

Read absorbance at 450 nm, and calculate amount of D-galacturonic acid released by comparing the OD of reaction to standard curve.

One unit of pectinase activity is defined as which liberates 1 μmole of D-galacturonic acid per minute at 37 °C, pH 5.0.

Calculate the value using Eq. 2 except T-C = Amount of reducing D-galacturonic acid liberated (Table 6).

Table 6
Standard curve of D-galacturonic acid

Concentration of D-galacturonic acid (μmol/ml)	Amount of reducing D-galacturonic acid stock solution (μl)	Amount of distilled water (ml)	Amount of reagent A (ml)	Amount of reagent B (ml)	Optical density at 450 nm
10	10	6.99	2		
20	20	6.98	2		
30	30	6.97	2		
40	40	6.96	2		
50	50	6.95	2		
60	60	6.94	2		
70	70	6.93	2		
80	80	6.92	2		
90	90	6.91	2		
100	100	0	2		
0 (blank)	0	1000	1		

Viva Voce:

Q1. What is the principle of enzyme assay?

Q2. Define specific activity and how it can be estimated.

Q3. Name substrates used for assaying protease activity.

Q4. Application of enzyme assays.

Chapter 18

Antimicrobial Sensitivity Assay

Materials: Petri plates, micropipette, sterilized L-shaped glass spreader, microtips, cork borer and filter paper, test pathogenic microorganisms, syringe filter Bunsen burner, 70% ethanol, sterilized cotton, ruler, permanent marker, water blank (10 ml distilled water and 2–3 drops of Tween 20 are mixed and autoclaved in screw cap tubes), Sabouraud dextrose agar (SDA) and SD broth, pH -7, glucose tryptone yeast extract broth, standard antibiotic disk/stock solution (e.g., penicillin 25 unit, 50 units, 100 units, 200 units, etc.), sterilized forceps.

Microorganism (test pathogen) can be procured from culture repositories like Microbial Type Culture Collection; Institute of Microbial Technology (IMTECH), Chandigarh; National Centre for Cell Science, Pune; and American Type Culture Collection Centre.

1 Procedure

Disk Diffusion Technique:

1. Pipette 0.1 ml of log phase culture of test pathogenic microorganism (as given in Table 1), or prepare suspension of pathogenic microorganism in sterilized water blank (2×10^4 cells per ml), and spread it over surface of pre-sterilized SDA media plates using L-shaped glass spreader.
2. The spreading should be done uniformly all over the plate.
3. Leave the plates for 5–10 min to dry surface of agar.
4. Place standard antibiotic disk/sterilized filter paper disk loaded by antibiotic equidistantly on the agar medium (Fig. 1.)
5. Incubate the plates at appropriate conditions as mentioned below:
   (a) Bacteria 37 °C for 24 h
   (b) Yeast 28 °C for 24–48 h

Aakanchha Jain et al. *Basic Techniques in Biochemistry, Microbiology and Molecular Biology: Principles and Techniques*, Springer Protocols Handbooks, https://doi.org/10.1007/978-1-4939-9861-6_18,

Table 1
Methods and requirements of antimicrobial sensitivity assay

S.no.	Technique/method	Media	Media composition	Observations
1	Disk diffusion technique	Sabouraud dextrose agar	Peptone 10.0 g, dextrose 20.0 g, agar 20.0 g, distilled water 1000 ml	Zone of growth inhibition around the disk
2	Minimum inhibitor concentration (MIC) assay	Sabouraud dextrose agar/broth	Peptone 10.0 g, dextrose 20.0 g, distilled water 1000 ml	Inhibition in growth of pathogenic microorganism, quantified using spectrophotometer
		Glucose tryptone yeast extract broth medium	Glucose 1.0 g, K_2HPO_4 1.0 g, yeast extract 5.0 g, tryptone 5.0 g, distilled water 1000 mL pH-8	

(c) Fungi 28 °C for 2–4 days

(d) Actinomycetes 28 °C for 2–4 days

6. Observe the plates for inhibition in growth of microorganism around the antibiotic disk.
7. Measure the diameter of growth inhibition using scale.
8. To prepare antibiotic disk in laboratory, make the standard solution of antibiotic (effective concentration) and filter sterilize it. Cut filter paper disk (8 mm, Whatman no. 2) using punching machine. Autoclave the filter paper disk (8 mm), and after cooling load the antibiotic (10 μl) on it under aseptic conditions; leave for 3–4 h to dry the antibiotic solution on disk.

Minimum Inhibitory Concentration: It is a quantitative method used for determining the minimum or lowest concentration (μg/ml) of an antibiotic which can cause inhibition in visible growth of microorganism in specified duration. It is used to make appropriate choice of antibiotic and to suggest the effective dosage of antibiotic under any treatment conditions. Also, antibiotic resistance can be prevented up to some extent.

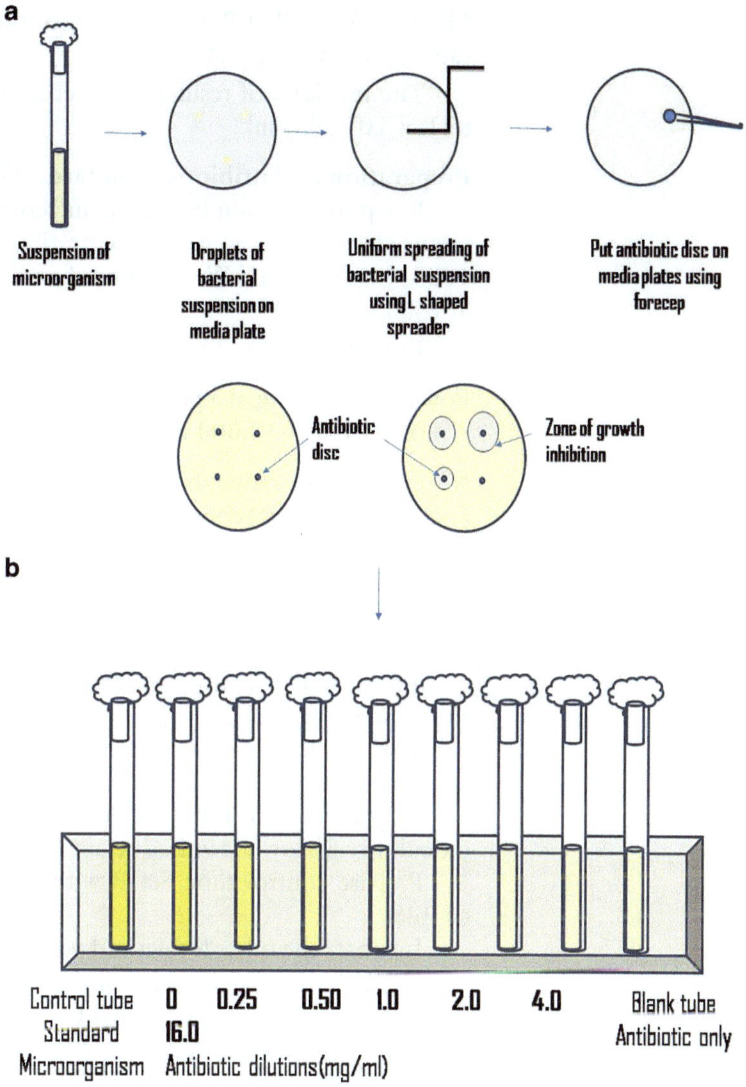

Fig. 1 Antibiotic sensitivity assay (**a**) Inhibition in growth of test pathogenic microorganism around the antibiotic disc; (**b**) Minimum Inhibitory concentration

2 Procedure

Preparation of Inoculum: The pure culture of microorganism is inoculated in glucose tryptone yeast extract broth medium and incubate for 4–5 h.

Preparation of McFarland Standard:

Part A: 0.5 ml of 0.048 M barium chloride (1.17% w/v).

Part B: 99.5 ml of 0.18 M sulfuric acid (1%v/v).

Mix Part A and Part B in screw-capped bottle, and tightly close the cap to prevent evaporation.

The turbidity of resultant suspension is considered equivalent to 1×10^6 cells/ml.

Preparation of Antibiotic Standard: $\frac{1000}{P} \times V \times C = W$.

P = potency given by the manufacturer (μg/mg), V = volume required (mL), C = final concentration of solution (multiples of 1000) (mg/ml), and W = weight of antibiotic (mg) to be dissolved in volume V (ml).

Example : $\frac{1000}{900} \times 20 \times 10 = 222.2\,\text{mg}$

Powder (222.2 mg at a potency of 900 μg/mg) dissolved in 20 ml of solvent = 10,000 mg/l.

Note: Antibiotic range should be prepared one step higher than the final dilution range required (Like 0 mg/ml, o.25 mg/ml, 0.5 mg/ml, 1.0 mg/ml, 2.0 mg/ml, 4 mg/ml, 8 mg/ml, 16 mg/ml, 32 mg/ml, 64 mg/ml, 128 mg/ml) so as to compensate for the addition of an equal volume of inoculum.

Filter sterilize antibiotic solution by syringe filter 0.22 μm.

Take 1 ml of antibiotic (concentration) in sterilized culture tubes/screw cap tubes.

To it add 1 ml of inoculum (equivalent density 1×10^6 cells/ml) and mix well.

Prepare control tube Set A with 1 ml of PBS and 1 ml of inoculum of known standard strain of microorganism.

Prepare control tube Set B with 1 ml of antibiotic and 1 ml of PBS.

Incubate the tubes for desired period at appropriate conditions.

Incubate the plates at appropriate conditions depending upon test microorganism as mentioned below:

(a) Bacteria: 37 °C for 24 h

(b) Yeast: 28 °C for 24–48 h

(c) Fungi: 28 °C for 2–4 days

(d) Actinomycetes: 28 °C for 2–4 days

Observe the tubes and read MIC end point as lowest concentration of antibiotic at which no visible growth is there.

The MIC of standard strain should be in one or two fold dilution of the expected MIC (already standardized).

Viva Voce:

Q1. Define antibiotics.

Q2. Name some antibacterial antibiotic.

Q3. Full form of MIC.

Q4. Why it is important to know MIC of antibiotic against microorganisms.

Chapter 19

Extraction of DNA from Plant Cells

Nucleic acids are polymeric macromolecules having monomeric nucleotides joined through phosphodiester bonds. The main composition consists of nitrogen bases (purines, adenine "A" and guanine "G"; pyrimidines, thymine "T," cytosine "'C," uracil "U"), sugars (oxyribose or deoxyribose), and phosphoric acid. Based on the type of sugar moiety, they are classified into two: ribonucleic acid (RNA) and deoxyribonucleic acid (DNA). Hydrolysis of these under stringent conditions yields nucleotide. The nucleotide on further hydrolysis gives nucleoside (sugar + bases) and phosphoric acid.

DNA and RNA have different structures. RNA has ribose sugar and base pair bonding A = U and G = C, while the sugar in DNA has one oxygen less at 3 position and base pairing A = T and G = C. Also, DNA consists of two polynucleotide chain present in highly supercoiled form, linked together through the above said bonds between two nucleotides. The bonding is in such a manner that the ratio of A/T and G/C is one. This is called as Chargaff's rule. The present structure of DNA was formulated by Watson and Crick in 1953. The DNA can be double stranded (ds) or single stranded (ss). On the other hand, Chargaff's rule is not applicable in RNA and that it is mostly single-stranded coiled structure.

DNA is made from genes, the unit of hereditary and variation. It is responsible for generation of all types of protein synthesizing signals of the body by the process of replication (formation of DNA from DNA), transcription (DNA to RNA), and translation (RNA to protein). It is found mostly in nucleus with traces in mitochondria and chloroplast in plants. It is also found in free form in many prokaryotes.

RNA is distributed throughout the cells, most of it cytoplasm and nearly 10% in nucleus and mitochondria. Five major types of RNA include rRNA (ribosomal), tRNA (transfer), mRNA (messenger), snRNA (small nuclear), and siRNA (small interfering). Some viruses possess double-stranded DNA.

Aakanchha Jain et al. *Basic Techniques in Biochemistry, Microbiology and Molecular Biology: Principles and Techniques*, Springer Protocols Handbooks, https://doi.org/10.1007/978-1-4939-9861-6_19,

1 From Onion

Principle: The tissue samples (leaves, flowers, onion, spinach, cauliflower) are mechanically broken using homogenizer or high-speed stirrers. Any proteolytic enzyme is used to remove protein contamination, and then the DNA is separated/concentrated using ice-cold ethanol.

Materials and Reagents: Homogenizer/blender, water bath, heating mantle, ice bucket, cheesecloth for filtering, onion, ice-cold ethanol (95%), homogenizing medium (liquid soap solution can also be used).

Homogenizing medium is comprised of:

(i) Sodium lauryl sulfate: 50 g [for protein denaturation in membrane].

(ii) Sodium chloride: 8.8 g [to maintain osmotic balance to avoid sudden cell rupture and DNA degradation].

(iii) Sodium citrate: 4.5 g [to maintain pH].

(iv) ethylenediaminetetraacetic acid: 0.3 g [for DNAse chelation]

Procedures

Preparation of Onion Extract:

- Cut the onion into small pieces. Place the pieces in a beaker containing homogenization medium or liquid detergent.
- Place the beaker in water bath and maintain the temperature to 60 °C for 15 min. Cool the solution instantly using an ice bath at 0–5 °C for 10–15 min with constant stirring.
- Pour the mixture into a homogenizer, and blend for only 5 s at high speed to degrade the cell wall membranes further, which permits the release of DNA.
- Filter the mixture in the second beaker. Ensure that the foam on the surface of the liquid should not contaminate the filtrate.

Separation of DNA from the Onion Extract:

- Add the enzyme (pepsin or papain or any proteolytic enzyme 0.01 M) to the onion tissue extract in a boiling tube, and mix well.
- Pour ice-cold ethanol from sides of tube on top of the onion extract.
- Leave the tube undisturbed for 2–3 min. As DNA is insoluble in ice-cold ethanol, it can only digest other compounds in the mixture leaving DNA as a precipitate.
- Gently rotate the glass rod in the liquid, at the interface of the alcohol and homogenized solution.

- The white web of mucus-like DNA can be drawn from the tube with the help of a Pasteur pipette, which is resuspended in 0.9% sodium chloride solution or stored in 50% ethanol.

Precaution:

- The ethanol used must be ice-cold; otherwise DNA will not be precipitated properly.
- The experiment must not be done by naked hands, wear gloves, as the DNAse enzyme from the hands may break DNA into small fragments and proper precipitate will not be obtained.
- Continuous stirring should be done. Stirring slows the breakdown of DNA at high temperature.
- Take care that blending should not be more than 5 s as it will degrade the DNA fiber.
- For proper spooling of DNA, the glass rod can be rubbed with sandpaper. This will attract the negatively charged DNA.

2 From Spinach and Cauliflower

Materials: Fresh plant tissue (fresh cauliflower heads).

5 M potassium acetate.

3 M sodium acetate.

70% ethanol.

Extraction buffer (TES buffer)

100 mM Tris-HC (pH 7.8)	6.06 g
10 mM disodium EDTA	9.30 g
500 mM NaCl	14.61 g
Water q.s to	500 ml

Suspension buffer

50 mM Tris-HCI (pH 8.0)	0.61 g
10 mM disodium EDTA	0.37 g
Water q.s. to	100 ml

TE buffer

10 mM Tris-HCl (pH 7.5)	0.12 g
10 mM disodium EDTA	0.37 g
Water q.s.	100 ml.

Procedure:

1. Grind 10 g of plant tissue to a fine powder.
2. Add 150 ml extraction buffer in small aliquots and grind thoroughly.
3. Transfer the hornogenate to a flask. Add 10 ml or 20% sodium dodecylsulphate [SDS], and mix thoroughly for 15 min using a magnetic stirrer.
4. Incubate the contents at 65 °C for 10 min.
5. Add 100 ml potassium acetate solution, mix, and incubate at 0'C for 30 minutes to precipitate polysaccharides and proteins.
6. Centrifuge at 25,000 g for 15 min and remove the precipitate.
7. Add six tenth volume (of supenatant) of isopropanol, and keep at 20 °C for 30 minutes to precipitate DNA.
8. Centrifuge again at 20,000 g for 15 min to obtain DNA pellet.
9. Discard the supernatant and drain off any liquid. Redissolve the pellet in suspension buffer (3 ml). Add 3.6 ml isopropanol and 360 ml 3 M sodium acetate solution, and keep at −20 ° C for 1 h.
10. Centrifuge again for 15 min (2000 g), wash with ice-cold ethanol (80%), and dry in vacuum.
11. Redissolve the DNA pellet in a suitable volume (1–5 ml) of TE buffer.

Chapter 20

Extraction of DNA from Animal Cells

1 From Blood [Using Phenol Extraction Method]

Principle: The proteins in a tissue homogenate are removed and precipitated at high salt concentrations by phenol and chloroform.

Reagents: Buffer I is composed of 10 ml of Tris-HCL (10 mM), 5 ml $MgCl_2$ (5 mM), and 0.32 M sucrose 109.5 g dissolved in 1000 ml of distilled water. Sometimes 10 ml (1%) Triton X–100 may be used to rupture cell wall.

Buffer II is made by mixing 50 ml EDTA pH 8.0 (25 mM) with 40 ml NaCl (75 mM) and adjusting the volume to 1000 ml.

Both the buffers are sterilized by autoclaving.

To prepare saturated NaCl solution, dissolve 2 g of NaCl in 100 ml water, and keep on adding 2–2 g till no more solid dissolves. Filter and autoclave.

Procedure:

1. Cool 10 ml of blood mixed with 9 ml of buffer I in ice bath for 5 min.
2. Centrifuge at 1500 rpm in a cooling centrifuge for 15 min; a pellet is formed.
3. Resuspend the pellet in 5 ml of buffer II to this add 0.5 ml of 10% sodium dodecylsulphate and 10 mg/ml, 55 μl proteinase K.
4. The mixture is kept overnight at 37 °C on a low-speed orbital shaker.
5. After this add saturated NaCI solution (approximately 1.4 ml) to each tube. Mix vigorously for a quarter of minute.
6. Centrifuge for 15 min at 2500 rpm. Precipitated protein pellet is obtained.

Aakanchha Jain et al. *Basic Techniques in Biochemistry, Microbiology and Molecular Biology: Principles and Techniques*, Springer Protocols Handbooks, https://doi.org/10.1007/978-1-4939-9861-6_20,

7. Transfer the supernatant to BD Falcon tube of nearly 15 ml capacity. Add exactly twice the volume of 95% ethyl alcohol, and slowly invert the tube several times until the DNA precipitate is seen.
8. Remove the DNA strands using a micropipette, and transfer to an Eppendorf tube containing 100–200 μl acetate buffer or TE buffer.
9. The DNA will dissolve. Use this solution for quantization using UV or any other method.

2 From Blood [Using Isopropanol Precipitation]

Reagents: For 50 ml lysis buffer, mix and dissolve 5 ml 100 mM Tris-HCl (pH 8.5) with 500 μl 0.5 M EDTA, 1 ml 10% SDS, and 2 ml NaCl (5 M) in distilled water. 0.25 ml proteinase K (20 mg/ml) is also mixed with it.

Procedure:

1. In a 100 ml flask, add ~6.7% of lysis buffer to blood sample ~20 ml. Incubate for 2–3 h at 37–40 °C stirring continuously. Filter.
2. Add one part of isopropanol to the lysate, and swirl it for 30 min to obtain precipitate.
3. The DNA is spooled using a micropipette. The DNA precipitate is dispersed in 0.5 ml, 10 mM Tris-HCI (250 μl), 0.1 mM EDTA (pH 7.5) (250 μl). Sometimes complete dissolution is obtained after 12–14 h incubation with continuous agitation.

3 From Mouse Liver

DNA isolation from an animal tissue sample depends largely on shear/vigorous shaking or vortexing and pipetting through a narrow pipette tip.

Procedure:

1. Fresh liver, weigh 0.5–1 g is obtained after dissection.
2. Crumble the liver in a clean plastic tray, into tiny masses. And keep them in pestle mortar.
3. Extract the contents using liquid nitrogen. Briefly, add liquid nitrogen (about 25 ml) to a mortar with liver pieces. Add extra 10 ml nitrogen when ~1/2 is left. Immediately start scrapping the liver chunks into the liquid, and leave it until no nitrogen is seen.

4. Powder the liver and add 5 ml digest buffer for thawing, stirring continuously.
5. When totally thawed, incubate the slurry in an oven at 50 °C for 12–14 h.
6. Add 15 ml phenol or chloroform to the 15 ml slurry containing liver homogenate, cap the tube, and mix thoroughly through hand. This forms a homogeneous mixture between the DNA phase and the phenol/chloroform phase.
7. Centrifuge at 4500 rpm for 5 min. Separate the upper aqueous layer using a micropipette in a new 15 ml tube.
8. Repeat steps 6 and 7.
9. Adjust the volume to 4 ml by either adding TE or removing DNA solution. Add 2 ml of 7.5 M ammonium acetate, and gently mix by inversion. Slowly add 8 ml of 100% ethanol by layering it over the DNA solution with a disposable pipette. Precipitate the DNA by slowly inverting the tube and rocking it back and forth. A threadlike DNA appears after ~ 3–5 min.
10. Centrifuge at 3000 rpm for 2 min. Discard the supernatant and wash the pellet thrice by 70% ethanol. Resuspend the pellet in 500 μl of TE.

4 From Pig Spleen

Principle: *Lymphoid organs and tissues (thymus) have high DNA percentage and low DNase, thus, facilitating high amount of DNA extraction. Further sodium citrate is used which chelates calcium and magnesium ion and does not allow the activation of DNase. The isolated deoxyribonucleoprotein is then solubilized in saline and deproteinized using chloroform/amyl alcohol. Ethanol is used for DNA precipitation.*

Materials and Reagents:

Pig spleen (300 g).

Solution A: For buffered saline (pH 7) [3 liters].

Sodium chloride 8.76 g (0.15 mol/liter) and sodium citrate 4.41 g/liter (0.015 mol/liter) in 1 liter distilled water.

Solution B: Sodium chloride 117 g in 1 liter distilled water (2 mol/liter).

Solution C: Chloroform/amyl alcohol (6:1) 2 liter.

2 liter of 95% ethanol.

500 ml diethyl ether.

Procedures:

1. Chop 50 g of pig spleen into chunks and homogenize for 1 min with 200 ml of solution A.
2. Centrifuge the homogenate for 15 min at 5000 g, again add 200 ml of buffered saline, and rehomogenize.
3. Separate the supernatant and discard. Suspend the sediment in 1 liter of solution B.
4. Centrifuge if any more sediments are seen. To the supernatant add distilled water to obtain 1:1 ratio.
5. Spool out the fibrous content (deoxyribonucleoprotein) using a sandpaper-rubbed glass rod. Allow to dry at room temperature.
6. Dissolve the contents in 100 ml of solution B, add 100 ml of solution C, and blend for 30 s.
7. Centrifuge for 10–15 min at 5000 g. The upper layer contains DNA. Care should be taken during separation because denatured proteins present at the interface of two liquids may contaminate the DNA.
8. Repeat step 7 to obtain at least 400 ml of liquid.
9. Add twice the volume of ice-cold ethanol to precipitate the DNA.
10. Remove the precipitate, and wash with grading concentrations (75, 80, 95%) of ethanol and finally with diethyl ether.

Chapter 21

Extraction of DNA and RNA from Microorganism

1 From Bacteria

Principle: The enzyme lysozyme is used for the rupturing of cell wall of bacteria or microorganism. The high salt concentration prevents the conversion of dsDNA to ssDNA. EDTA and sodium citrate present helps to chelate metal ions which activate DNase and RNase enzymes. The protein denaturation is done by sodium chlorate and finally extraction is done either by ethanol or isopropanol method as discussed earlier in section.

Material and Reagents:

Any bacteria (for the first timers, use *Escherichia coli*).

Solution A: Saline EDTA pH 8.0 in distilled water.

Sodium chloride (0.15 mol/liter) = 125 ml.

EDTA (0.1 mol/liter) =125 ml.

Enzyme lysozyme (N-acetylmuramide glycanhydrolase) = 100 mg.

Sodium lauryl sulfate (10% w/v) = 50 ml.

Sodium chlorate (6 mol/liter) = 100 ml.

Chloroform: Isoamyl alcohol (24: 1) = 500 ml.

Absolute ethanol.

Sodium acetate (3 mol/liter) = 25 ml.

Isopropanol = 100 ml.

Sodium hydroxide (6 mol/liter) = 50 ml.

Aakanchha Jain et al. *Basic Techniques in Biochemistry, Microbiology and Molecular Biology: Principles and Techniques*, Springer Protocols Handbooks, https://doi.org/10.1007/978-1-4939-9861-6_21,

Solution B: Saline citrate buffer (SSC × 10) pH 7.0.

Sodium chloride (0.15 mol/liter) mixed with sodium citrate (0.15 mol/liter) in distilled water.

Procedure:

1. Mix 3 g of cells with 30 ml of solution A.
2. Add 15 mg N-acetylmuramide glycanhydrolase to the above solution, and stir at 37 °C for 20 min.
3. Add 6 ml of sodium lauryl sulfate and stir at 60 °C for 10 min. Cool it suddenly on ice bath till 37 °C and filter.
4. Add 7 ml of sodium chlorate stirring continuously. Further add 40 ml of chloroform: isoamyl alcohol. After complete addition, shake vigorously for 15–20 min.
5. The DNA can now be precipitated by either ethanol or isopropanol as narrated in experiment 2.3 (B) ((a) and (b)).
6. Dissolve the DNA in 10 ml of solution B and add RNase (100 μg/ml); keep at 37 °C for 30 min.
7. Add 2 ml of sodium chlorate and 15 ml of chloroform: isoamyl alcohol; shake slowly for further 15 min.
8. Centrifuge at 10000 g for 5 min.
9. Precipitate the DNA using ice-cold ethanol.
10. Dissolve the water-removed DNA in solution B.

2 From Yeast Cells

Principle: The nucleic acid-protein complex in a cell (microbe) is dissociated into RNA, and protein is removed using phenol. The RNA left is precipitated using alcohol.

Materials: Yeast tablets (200 g), phenol (900 g/L redistilled), potassium acetate (20% w/v; pH 5; 200 mL), absolute ethanol (3 L), diethyl ether (500 ml). All the glassware must be sterilized before use.

Procedure:

- Activate the yeast powder by suspending it in warm water ~ 40 °C. The process requires nearly 15–20 min.
- Freshly redistilled phenol is added and stirred for about 30 min.
- The solution is then centrifuged for 15 min at 3000 rpm. The two layers will be separated; the phenolic layer has DNA, while the aqueous upper layer contains RNA, carbohydrates, etc.
- Separate the aqueous layer and filter using glass wool.

- Again centrifuge for 5 min at 10,000 g in a cooling centrifuge. Add potassium acetate at a concentration to have the final approximation of 2% w/v.
- Add two volumes of ethanol; keep at -20 °C for RNA precipitation (approx. 1 h).
- The precipitate of RNA is separated by centrifugation at 2000 g for 10 min.
- Wash the pellet (RNA) with 70% ethanol, then with ethanol-ether (1:1), and finally with ether.
- Dry the pellet in vacuum.

Note: Sometimes carbohydrates (polysaccharides) contaminants may also be present.

Viva Voce:

Q1. What is DNA? Differentiate DNA and RNA.

Q2. Why is it important to wear gloves during experiment?

Q3. Give the composition of homogenizing medium.

Q4. What is the purpose of each ingredient added in homogenizing medium?

Q5. What is the use of ice-cold ethanol?

Q6. What is the charge on DNA?

Q7. Why are the homogenizing medium and onion extract cooled suddenly after heating?

Q8. Which enzymes are used to remove contaminating proteins during extraction?

Q9. How is DNA stored after extraction for further use?

Q10. Who discovered double stranded structure of DNA?

Chapter 22

Estimation of DNA by Diphenylamine Reaction

Principle: In the presence of acid, the 2-deoxypentose of DNA is converted to beta-hydroxylevulinaldehyde which reacts with diphenylamine to give blue color, which is estimated by UV spectrophotometer or by colorimeter at 595 nm. Since only deoxyribose of purine nucleotide reacts, the value obtained represents only half to total amount.

Reagents:

1. DNA standard (100 mg/ml).
2. Standard citrate saline buffer (0.15 M sodium chloride +0.015 M sodium citrate).
3. For diphenylamine reagent (DPAR), dissolve 1.5 g diphenylamine in 100 ml of acetic acid, and add 2.75 ml of concentrated sulfuric acid. Store the reagent at 2 °C, and use within 6 months.

Procedures:

1. Prepare 100 mg/ml DNA standard solution and unknown isolated DNA sample in citrate buffer.
2. Prepare 3 ml aliquots of DNA standard (50–500 μg/ml).
3. Add 6 ml of DPAR to each aliquot. Mix properly and heat the tubes in boiling water bath for 10 min and cool.
4. Using citrate buffer as blank, take the absorbance at 595 nm by UV spectrophotometer or by colorimeter.
5. Take the absorbance of unknown sample = AA (Table 1).

Aakanchha Jain et al. *Basic Techniques in Biochemistry, Microbiology and Molecular Biology: Principles and Techniques*, Springer Protocols Handbooks, https://doi.org/10.1007/978-1-4939-9861-6_22,

Table 1
Standard curve of DNA

S.no.	Concentration of DNA (μg/ml)	Absorbance (nm)
1.	50	
2.	100	
3.	150	
4.	200	
5.	250	
6.	300	
7.	350	
8.	400	
9.	450	
10.	500	

Calculations
Obtain the equation $y = mx + c$ from the above data. Calculate the values of × for AA absorbance, i.e., unknown sample

Applications:

For Estimation of DNA Content In

(i) Samples obtained from new species of plants and animals

(ii) Samples obtained from cell line studies after defragmentation, apoptosis, etc.

(iii) In vivo samples obtained after new drugs given for treatment of diseases like cancer and autoimmune diseases

(iv) The study of the effect of mutagens on DNA content in organisms

Viva Voce:

Q1. Give the full form of DNA.

Q2. What are the major differences between DNA and RNA?

Q3. What is the principle of estimation of DNA by diphenylamine reaction?

Q4. Which color is obtained from diphenylamine reaction?

Q5. Give the composition of citrate saline buffer.

Q6. Give the composition of diphenylamine reagent and its validity.

Q7. What are the applications of DNA estimation?

Q8. The DNA gives which color with diphenylamine reagent, and tell the lambda max of its estimation.

Q9. Name the instruments used to quantify the color obtained after diphenylamine reaction.

Q10. Which part of DNA gives color with diphenylamine reagent?

Chapter 23

Estimation of RNA Using Orcinol Method

Principle: Pentose sugar present in ribonucleic acid reacts with hot acid to form furfural, which on reaction with orcinol and ferric chloride produces green color. Purine nucleotides are more reactive than pyrimidine. The green color can be quantitated using UV spectrophotometer or colorimeter at 665 nm.

Materials and Reagents:

1. Standard RNA solution or pure RNA extracted from yeast.
2. Sample/unknown RNA.
3. Orcinol solution (6% w/v in 95% ethanol) [stored in cool place].
4. Bial's reagent/orcinol acid reagent (10% w/v ferric chloride hydrated 1 ml added to concentrated hydrochloric acid 200 ml).
5. Tris-acetate buffer (1 mM EDTA added to 10 mM tris-acetate, pH 7.2) or any buffer of pH 7.2.

Other color reactions of RNA are with aniline and phloroglucinol.

Procedures:

1. Dissolve 100 mg RNA pure standard in ice-cold tris-acetate buffer.
2. Prepare 3 ml aliquots of RNA standard (50–500 μg/ml) using RNAse-free water.
3. Add 6 ml of orcinol acid reagent to each aliquot. Mix and add 0.4 ml of orcinol solution. Mix again.
4. Heat the tubes in boiling water bath for ~20 minutes and cool.

Aakanchha Jain et al. *Basic Techniques in Biochemistry, Microbiology and Molecular Biology: Principles and Techniques*, Springer Protocols Handbooks, https://doi.org/10.1007/978-1-4939-9861-6_23,

5. 3 ml RNAse-free water with all other ingredients in steps 3 following step 4 were used as blank.
6. The absorbance was read at 665 nm by UV spectrophotometer or colorimeter (Table 1).
7. Take the absorbance of unknown sample = RR.
8. Calculations.
9. Obtain the equation y = mx + c from the above data. Calculate the values of x for RR absorbance, i.e., unknown sample.

Applications: For estimation of RNA content in:

(i) Samples obtained from new species of plants and animals
(ii) Samples obtained from cell line studies after gene modification to produce novel proteins
(iii) In vivo samples obtained after new drugs given for treatment of diseases

Viva Voce:

Q1. What are nucleic acids? Give the full form of RNA.
Q2. Which part of RNA gives color with orcinol reagent?
Q3. What is the principle of estimation of RNA by orcinol reaction?
Q4. Which color is obtained from orcinol reaction?

Table 1
Standard curve of RNA

S.no.	Concentration of RNA (μg/ml)	Absorbance (nm)
1.	50	
2.	100	
3.	150	
4.	200	
5.	250	
6.	300	
7.	350	
8.	400	
9.	450	
10.	500	

Q5. Give the composition of tris-acetate buffer.

Q6. Give the composition of orcinol reagent.

Q7. What are the applications of RNA estimation?

Q8. The RNA gives which color with orcinol reagent, and tell the lambda max of its estimation.

Q9. What is the other name of orcinol reagent?

Q10. Name the other reactions used for RNA estimation.

Chapter 24

Determination of Melting Temperature of DNA

Principle: Heat causes the disordering of double helical structure of double-stranded (ds) DNA. This causes the breaking of double and triple bonds between two nucleotides in the complementary strands of DNA. The temperature at which this occurs is called as helix-to-coil transition temperature or melting temperature of DNA "T_m." The separated single-stranded DNA (ssDNA) transmits 60% less ultraviolet light as compared to the dsDNA at 260 nm. The hyperchromic shift is used to analyze the melting profiles of DNA. The cooling curve can also be obtained by cooling the whole solution slowly.

Materials and Reagents:

1. Deoxyribonucleic acid (DNA).
2. Ribonucleic acid (RNA).
3. 0.5 mol/liter acetate buffer (pH 5.5)
4. Buffered saline solution (0.15 mol/liter sodium chloride).
5. Buffered saline solution (0.105 mol/liter sodium citrate; pH 7.0).
6. Water bath.

Procedures:

1. Prepare three solutions of DNA in different strength buffered saline to get 100 μg/ml solution.
2. The change in extinction or absorbance at 260 nm in all the three tubes is recorded separately by continuously increasing the temperature from 0 °C to 90 °C. Plot the graph or melting curve taking extinction at y-axis and temperature at x-axis.

Aakanchha Jain et al. *Basic Techniques in Biochemistry, Microbiology and Molecular Biology: Principles and Techniques*, Springer Protocols Handbooks, https://doi.org/10.1007/978-1-4939-9861-6_24,

3. Repeat the step 2 but in reverse direction, i.e., from 90 °C to 0 °C. Plot an overlay graph or cooling curve over melting curve (Fig. 1) taking extinction at y-axis and temperature at x-axis.
4. To study the influence of ionic strength on DNA, repeat steps 2 and 3 for each strength of DNA solution prepared (Table 1).

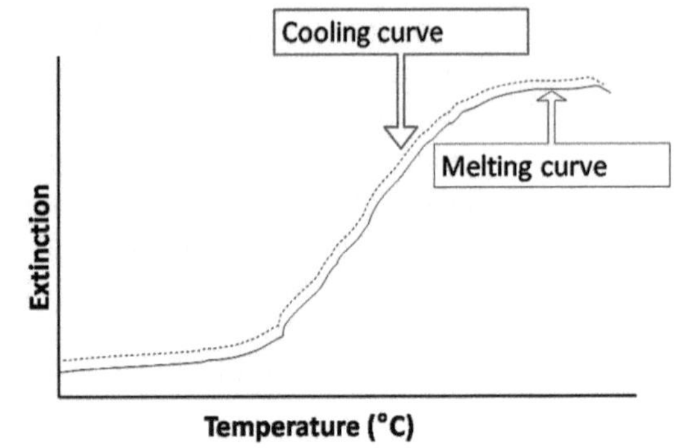

Fig. 1 Plot of an overlay graph of cooling curve over melting curve

Table 1
Extinction of DNA cooling and heating cycle measured at 260 nm

S.No.	Extinction at 260nm					
	Dilute DNA solution (0.1)		DNA solution in 0.15mol/liter sodium chloride)		DNA solution in 0.105mol/liter sodium citrate; pH 7.0	
	Heating	Cooling	Heating	Cooling	Heating	Cooling

Chapter 25

Quantitative Estimation of DNA and RNA

Principle: The purine (adenine and guanine) and pyrimidine (cytosine, thymine, and uracil) bases of DNA and RNA can be estimated quantitatively and qualitatively using UV spectrophotometer, because the conjugated double bonds of nucleic acids tend to absorb ultraviolet light between 220 nm and 300 nm. On solubilizing the DNA in citrate buffer and measuring its optical density at 260:280 nm, the ratio of 1.85 indicates that the isolated DNA is pure. The ratio changes to ~2.0 for RNA and decreases below 1.85 for protein contaminations if any. Nomogram is the scale which indicates the amount of both RNA and DNA in a sample at the same time.

Combined *Viva Voce* for Experiment Determination of Melting Temperature of DNA and Quantitative Estimation of DNA and RNA:

Q1. What is meant by melting of DNA?

Q2. What is the principle of determining melting temperature of DNA?

Q3. Which buffer is used for dissolving DNA?

Q4. What is the most abundant form of DNA in nature?

Q5. ____________temperature range is used for determining melting temperature of DNA.

Q6. What is the absorbance ratio of pure DNA at 260:280 nm?

Q7. The value of 2.0 obtained for 260:280 ratio at UV is for____________

Q8. The value less than 1.85 indicates____________contamination in isolated DNA sample.

Q9. The scale used for estimation of RNA and DNA in a sample at a time is called as____________

Q10. Which wavelength is used for UV estimation of DNA and RNA?

Aakanchha Jain et al. *Basic Techniques in Biochemistry, Microbiology and Molecular Biology: Principles and Techniques*, Springer Protocols Handbooks, https://doi.org/10.1007/978-1-4939-9861-6_25,

Chapter 26

Purification and Bioassay of Interleukin-1 and Interleukin-2 (IL-1 and IL-2)

The term interleukin derives from (inter-) "as a means of communication" and (-leukin) "deriving from the fact that many of these proteins are produced by leukocytes and act on leukocytes." Cytokines are involved in amazingly broad array of biological activities including innate immunity, adaptive immunity, inflammation, and hematopoiesis. They can modulate tumor growth and microenvironment by mediating interactions between cancer cells and infiltrating inflammatory cells. On the basis of their function, cytokines can be grouped into anti- or pro-inflammatory cytokines, chemokines, and growth factors. IL-1 and IL-2 play a crucial role in the T-cell activation process. IL-1, interleukin-1, is also called as lymphocyte-activating factor, mitogenic protein, T-cell-replacing factor III, B-cell-activating factor, B-cell differentiation factor, and "Heidikine." IL-1 alpha and IL-1 beta are the two forms of IL-1 having molecular weight of 17 KD each and isoelectric points of 5 and 7, respectively. The procedure for bioassay of both is same.

Interleukine-1 (IL-1): Principle: The assay is based on the capability of IL-1 to produce T_H cell by thymus. Also increased DNA synthesis depends on the concentration of IL-1.

Requirements:

Mouse: BALB/C strain.

Medium: RPMI 1640 supplemented with L-glutamine, 0.2% sodium bicarbonate, HEPES buffer (25 mM), gentamycin (50 μg/ml), penicillin (40 IU), streptomycin (100 μg/ml), fetal calf serum [FCS] (10–12%) at pH 7.2 ± 1.

Siliconized Pasteur pipettes, flask, centrifuge tubes, Petri dishes, 96-well microtiter plate, cell harvester.

Aakanchha Jain et al. *Basic Techniques in Biochemistry, Microbiology and Molecular Biology: Principles and Techniques*, Springer Protocols Handbooks, https://doi.org/10.1007/978-1-4939-9861-6_26,

Purification of IL-1: IL-1 can be generated by stimulating cells of macrophages lineage. Mouse peritoneal cells or some tumor cell lines are used for this purpose. Cells are cultured (1–2 × 10^6cells/ml) in RPMI 1640 medium with 10% FCS and 50 μg/ml gentamycin, with activating agents like bacterial polysaccharide (LPS) for 1–4 days. IL-1 can be purified from the culture supernatant.

Bioassay:

(I) Preparation of Mouse Thymocytes

- Mice are dissected after cervical dislocation and thymus is removed in 5 ml medium.
- Rinse and transfer to fresh 5 ml medium.
- Squeeze thymocyte and keep allow the coarse particles to settle.
- Centrifuge the supernatant at 250 g for 5 min.
- Wash the sediment two to three times with 5 ml of complete RPMI medium.
- Adjust the cell suspension culture concentration to 15 × 10^6 cells/ml.

(II) Production of IL-1 by Macrophages

1. Collect the mouse peritoneal exudates cells under aseptic condition, and suspend in incomplete medium in siliconized glassware (helps in removal of adhering macrophages).
2. Adjust the concentration of macrophages to 1 × 10^6 cells/ml in complete RPMI medium.
3. Transfer 5 ml of this culture to Petri dish and incubate for 2 h at 37 °C.
4. Wash the cells in incomplete medium and resuspend in 5 ml of complete medium.
5. Incubate this culture at 37 °C in humid chamber at 5% CO_2 for 4 days.
6. After incubation, centrifuge at 6000 g for 10 min, and collect the supernatant which contains IL-1.

(III) Assay: The test is done in quadruplicate.

Thymus control: 0.1 ml cell approx. 15 × 10^6 cells /ml; 0.1 ml control medium.

Phytohemagglutinin (PHA) control: 0.1 ml cell; 0.1 ml medium; 10 μl PHA (2 μg/ ml).

Test: 0.1 ml cell; 0.1 ml culture supernatant or IL-1 standard; 10 μl PHA.

1. Incubate the plates in humid chamber at 37 °C for 3 days.
2. After incubation add 10 ml of radioactive thymidine 3H (0.5 μci) and again incubate overnight.
3. Wash the cells next day with cold thymidine solution.
4. Dissolve the pellet obtained in 1 N NaOH.
5. Add 10 ml of scintillation fluid.

$$1\text{U of IL 1} = \text{Volume of solution per reciprocal of titer value which results in 50\% incorporation of 3H thymidine.} \quad (1)$$

Interleukin-2 (IL-2): Principle: IL-2 is produced by T lymphocytes and stimulates production of other lymphokines and upgrades IL-2 receptors expression. Interleukin-2 (IL-2, formerly termed T-cell growth factor), discovered by Morgan et al. (1), is produced by T lymphocytes after antigen or mitogen stimulation and is required for the proliferation of activated T cells. This enhances function such as phagocytosis, cell killing, and antibody production. Cell line which elicits mitogenic responses to IL-2 forms the basis of this assay.

$$1\text{U of IL2} = \text{Volume of solution per reciprocal titer which results in 50\% incorporation of 3H thymidine in IL2 – sensitive cell lines (CTLL 2)} \quad (2)$$

Materials:

- CTLL-2 cells harvested on third day after IL-2 feeding.
- 96-well microtiter plate.
- Growth medium: RPMI 1640 containing 5×10^{-5}M mercapto-ethanol, 10% FCS, 100 units penicillin/ml, and 150 μg streptomycin.
- IL-2 standard and control.

Partial Purification of IL-2:

1. Mouse spleen cells are adjusted to 5×10^{6} cells/ml in RPMI medium supplemented with 2×10^{-5}M mercapto-ethanol, 2% FCS, and 50 μg/ ml streptomycin.
2. Cell suspensions are incubated with optimal concentration of mitogen (PHA for human cells) for 2 days in CO_2 incubator.

3. Cells are centrifuged and proteins in clear supernatant are precipitated with ammonium sulfate.
4. Centrifuge for 20 min at 10000 rpm, and dissolve in minimum quantity of normal saline.
5. The material obtained is dialyzed against ice-cold phosphate buffer saline to remove excess ammonium sulfate.
6. Fractionate the clear brown solution on Sephadex G-100 column at 4 °C.
7. Each column filtrate is filter sterilized (0.45 μ) and test for IL-2 activity is done.

Bioassay:

(A) Maintenance of CTLL-2 Cells

Cloned mouse cytotoxic cell line is maintained in RPMI 1640 medium containing 5% FCS, 100 unit penicillin/ml, 100 μg streptomycin, and appropriate dilution of IL-2. The cells are grown in stationary suspension.

(B) Production of IL-2

IL-2 can be obtained from fresh tissue or from cell lines of human peripheral blood leucocytes enriched on Ficoll Hypaque density gradient. PHA (10 ng/ml) may increase yield, but because of carcinogenicity, it may interfere in the assay.

(C) Assay

1. Harvest 4000 murine IL-2-dependent cytotoxic T lymphocyte cells (CTLL) in 96-well microtiter plates, so that each well has 0.2 ml of total volume.
2. Incubate for 24 h.
3. Ass 0.5 μCi of tritiated thymidine [3H]TdR (20 Ci/mmol) to each well. Incubate for next 4 h.
4. Harvest the cells on glass fiber strips and [3H]dT incorporation measured in a liquid scintillation counter.
5. 1 U/ml of IL-2 was defined as the quantity of IL-2 released in 48 h culture medium conditioned by rat spleen cells (1 × 106/ml) stimulated by concanavalin A (Con A; 5 μg/ml).

Viva Voce:

Q1. What is the principle for production of IL-1?

Q2. Which medium is used for growing cell line CTLL?

Q3. Interleukin-2 was formerly known as ______________

Q4. Interleukin-1 is also called as ______________

Q5. 1 IU of IL-1 = ______________________________

Q6. 1 U of IL-2 = ______________________________

Q7. State whether true/false.

(a) IL-2 is an essential mediator of the immune response______________

(b) IL-2 is responsible for the abnormal cell proliferation in human lymphoblastic leukemias______________

Q8. Define interleukin.

Q9. What were the other names of IL-1?

Chapter 27

Quantitative Determination of TNF Alpha and Interleukin-6

Cytokines are key regulators of immune responses in cancer. Cytokines are involved in amazingly broad array of biological activities including innate immunity, adaptive immunity, inflammation, and hematopoiesis. In addition to cytotoxic T cells, there are **helper T cells (T_H cells)**, whose function is to produce soluble signaling proteins called cytokines, which include the interleukins that stimulate T_C, T_H, and B cells to proliferate, activate, and stimulate different classes of leukocytes and possible uses in treatment of wounds, HIV infection, cancer, and immune deficiencies. Interleukin (IL) is one of several proteins important for lymphocyte proliferation. IL-1 is produced by macrophages and induces the production of IL-2 by T cells that have been stimulated by antigen or mitogen. IL-2 stimulates the production of T cells bearing specific receptor for IL-2; these receptors are expressed in response to the antigenic stimulation. IL-2 also seems to induce the production of interferon and is used as an anticancer drug in the treatment of wide variety of solid malignant tumors. Another interleukin, IL-3, is necessary for the differentiation of suppressor T cells. They can modulate tumor growth and microenvironment by mediating interactions between cancer cells and infiltrating inflammatory cells. On the basis of their function, cytokines can be grouped into anti- or pro-inflammatory cytokines, chemokines, and growth factors. Moreover, the classification into Th1 and Th2 cytokines is of importance. In general, Th1-type cytokines, e.g., interleukin (IL-12, IL-15) and interferon ɣ (IFN- γ), contribute to cellular immune reactions considered essential for an effective response against tumor cells, whereas Th2-type cytokines, e.g., IL-4, IL-6, IL-10, and IL-13, may suppress the tumor-specific immune response.

TNF alpha is a potent immunoprotein produced in response to tumor infection by B and T lymphocytes, macrophages, or any transformed cells (e.g., cancerous cells). Immune system cells associated with tumors, called tumor-infiltrating lymphocytes, can be

Aakanchha Jain et al. *Basic Techniques in Biochemistry, Microbiology and Molecular Biology: Principles and Techniques*, Springer Protocols Handbooks, https://doi.org/10.1007/978-1-4939-9861-6_27,

genetically modified to produce tumor necrosis factor (TNF). When these lymphocytes are taken from a cancer patient, modified, and reintroduced, the engineered cells target the tumor, and they produce TNF which causes tumor shrinkage. The regulatory mechanisms that trigger apoptosis involve some of the same proteins that regulate the cell cycle. The signal for suicide often comes from outside, through a surface receptor. TNF, produced by cells of the immune system, interacts with cells through specific TNF receptors. These receptors have TNF-binding sites on the outer face of the plasma membrane and a "death domain" of about 80 amino acid residues that passes the self-destruct signal through the membrane to cytosolic proteins such as TRADD (TNF receptor-associated death domain). Apoptosis can be triggered by extracellular signals such as TNF through plasma membrane receptors.

Principle: *Enzyme-linked immunosorbent assay (kits available in the market).*

Reagent Preparation:

1. Working detector (50 mL).
   10 mL of the detection antibody "biotin antihuman TNF alpha" is mixed with 40 μL of enzyme concentrate (streptavidin-horseradish peroxidases).
2. Wash buffer is 20X concentrated phosphate-buffered saline (pH 7.2) with Tween 20 (20 X 100 mL) = 5 mL buffer in 95 mL double distilled water (DW).

Procedure:

1. Collect the serum samples from animals of all the groups for cytokines analysis.
2. Take a 96-well ELISA plate.
3. The wells coated with specific capture antibody are incubated overnight at 4 °C.
4. Coated wells are blocked with assay diluents (50 μL) and incubated at room temperature (RT) for 1 h.
5. Add 100 μL standard/ sample diluents as a negative control (available in kit).
6. Add 100 μL of standard as a positive control.
7. The serum samples were added (100 μL) to their respective wells and incubated for 2 h at RT.
8. Wash the wells by 300 μL of wash buffer five times.

9. Prepare the working detector (detection antibody + conjugate) 15 minutes before this step.
10. Add 100 µL of working detector to each well.
11. Incubate both the well for 1 h at room temperature (25 ± 1 °C).
12. Wash seven times with 300 µL wash buffer.
13. Add 100 µL of TMB (substrate) to each well.
14. Incubate at room temperature for 30 minutes.
15. Add 50 µL of stop solution in each well.
16. Take reading at 450 nm in ELISA reader.

Applications:

1. For detecting the change in the cytokine levels in HIV infection, cancer, and immune deficiencies
2. For analyzing the process of apoptosis

Viva Voce:

Q1. What are interleukins?

Q2. Define monocytes and macrophages. How do they differ?

Q3. What is the composition of wash buffer?

Q4. Which substrate is used in the assay?

Q5. At what lambda max ELISA reading is done?

Q6. Explain the protocol for IL assay.

Q7. What is cancer and how is it related to IL?

Q8. What are the different cells of immune system?

Q9. State whether true or false:

(a) IL activates and stimulates different classes of leukocytes.…… ..

(b) IL is used in treatment of wounds, HIV infection, cancer, and immune deficiencies.………… .

Part III

Microbiology

Chapter 28

Sterilization of Glassware; Preparation and Sterilization of Media

Sterilization (Latin *sterilis* meaning barren) is the complete removal or inhibition of all live forms (living cell, spores, virus, etc.) from any object or place. Petri plates, flasks, beakers, pipettes, etc. are some of the generally required glassware in microbiological laboratories. Sterilization of glassware by moist heat using autoclaving is not very effective since moisture enters the glassware and may cause contamination problems. The most effective and used method of sterilization of glassware is dry heat treatment using hot air oven. On exposing the glassware to dry heat, the inhibition in growth occurs due to denaturation of proteins and oxidation of biomolecules. Duration of sterilization is temperature dependent: at higher temperature sterilization is achieved faster, i.e., at 170 °C, 30 min; 160 °C, 60 min; and 150 °C, 150 min or longer. Spores of *Bacillus atrophaeus* are used as biological indicator to test the efficacy of sterilization process. Since glasswares are sensitive to rapid changes in temperature, they are wrapped in paper and then only kept in hot air oven.

Materials: Newspaper, thread, Petri plates, beaker, pipette, hot air oven, spore strip of *Bacillus atrophaeus*.

Procedure:

Petri Plate Sterilization:

1. Take clean Petri plates and allow it to dry at room temperature.
2. Make a bundle of 12 Petri plates (upper lid and lower lid) by keeping the plates one above another; arrange the plate such that lower lid is toward base and upper lid covers it.
3. Tear newspaper/brown paper, to match the width of 12 plates, keep on the table to the width position, and keep the bundle of Petri plate at one edge (Fig. 1).

Aakanchha Jain et al. *Basic Techniques in Biochemistry, Microbiology and Molecular Biology: Principles and Techniques*, Springer Protocols Handbooks, https://doi.org/10.1007/978-1-4939-9861-6_28,

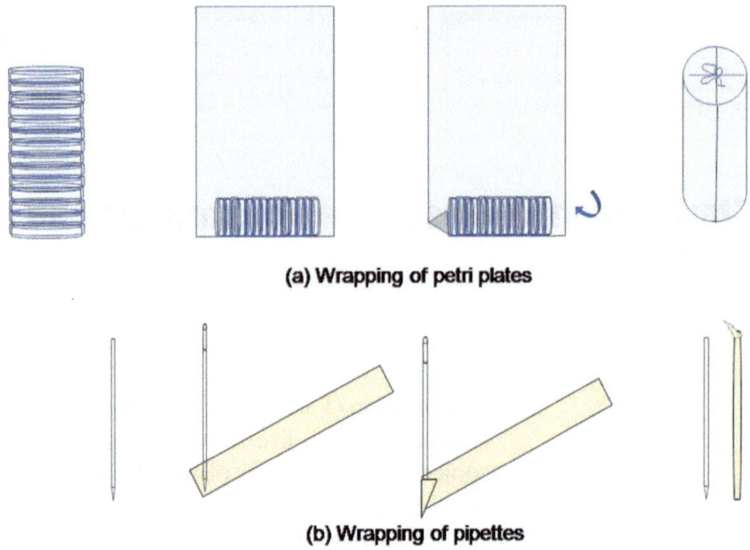

Fig. 1 Wrapping of petri-plate and pipette for sterilization

4. With the help of the thumb, the newspaper is folded over the bundle of plate; the folding is continued with rotation to cover the circumference of the bundle. Check that the paper covers entire bundle.
5. Tie thread over it in crossed manner to avoid opening of paper around the bundle. Mark directions and mention date and time on bundle.
6. Keep in hot air oven at 160 °C for 1–2 h.

Pipette Sterilization:

1. Take clean glass pipette and put cotton beak on its broader opening.
2. Cut a strip of paper 15 cm × 5 cm long.
3. Keep the narrower edge of pipette at an angle of 30 ° for the paper strip.
4. Fold one edge tightly and then roll paper over pipette till its other edge.
5. Screw and fold the paper at the other edge. Remove extra paper. Mention date and time of wrapping.
6. Keep in hot air oven at 160 °C for 1–2 h.

Testing of Sterility:

1. Wrap the spore strip of *Bacillus atrophaeus/Bacillus subtilis* in paper envelop, and keep in oven for desired duration.

2. After completion of sterilization process, take out the envelop of spore strip, and transfer it on nutrient agar media plate under aseptic condition, and incubate the plates at 37 °C for 24 hours.
3. Observation: No growth of *Bacillus atrophaeus/Bacillus subtilis* on plate confirms sterilization.

Precaution:

1. Switch on the oven at least 1 h before keeping the glassware.
2. Check the temperature of oven with thermometer. When exact temperature reaches as desired, then only keep the glassware inside it.
3. Take out glassware 1 h before use.

Culture Medium: Like any other living organism, microorganisms require nutrients for their growth and survival. The general growth requirement of any microbial growth support medium is source of carbon, nitrogen, phosphorus, sulfur, trace elements, and water. Under laboratory condition the nutrient mixture on which microorganisms grow is called culture medium, and microorganism is termed culture. Since the growth requirement of different microorganisms differs greatly, thus choice of medium to be used is dependent of the type of microorganism and its purpose. The important thing to note is that the process of media preparation is the most important part of any culturing process since improper handling or preparation of media may cause loss of nutrient value that will further cause adverse effect on growth of microorganism. The requirement of nutrient for bacteria, fungi, and actinomycetes differs greatly. Most of the bacteria and fungi require organic source of nutrient for their proper growth, while actinomycetes prefer to grow on inorganic nutrients. Compositions of different types of culture media generally used for bacteria, fungi, and actinomycetes are described in Tables 1 and 2 and Appendix 1.

Materials: Petri plates, filter paper, Bunsen burner, 70% ethanol, sterilized cotton, Tween 80, inoculation loop, medium ingredient as mentioned in Table 2, culture tubes, laminar airflow, distilled water, glass screw cap tubes, vortex mixture, conical flask.

Procedure:

1. Take clean, dry 500 ml of conical flask; prepare cotton plug for mouth of flask; the plug should be tight enough to restrict passage of any microbial contaminant and loose enough to allow exchange of air and moisture inside the flask.
2. Calculate the composition of medium for 300 ml.

Table 1
Different types of medium

S. No.	Basis of classification of media	Types of medium	Properties	Example
1	Physical state	Solid culture media	Agar is added to medium as solidifying agent	Nutrient agar potato dextrose agar
		Broth culture media	Without agar	Lactose broth
		Semisolid culture media	Semisolid nature of ingredients	Soft nutrient agar
2	Media composition	Natural or empirical culture media	Exact composition of not known	Blood agar meat extract medium
		Semisynthetic media	Composition is partially known	Potato dextrose agar nutrient agar
		Synthetic or chemically defined media	Composition is known	Czapek dox medium Raulin's medium
		(a) General purpose medium		Richard's medium
		(b) Selective medium	Medium ingredient allows growth of specific microorganism and inhibit growth of others	Mannitol salt agar
		(c) Differential medium	Medium ingredient causes different colonies to grow differently	Blood agar
		(d) Assay medium	Contains substrate or compound that with itself indicated the presence of bioactive compound or will react with product formed to give qualitative result of assay	Czapek dox medium supplemented with asparagine (asparaginase assay medium)

Table 2
Growth medium for different types of microorganisms

S. No.	Microorganism	Media name and composition[a]
1	Bacteria	Nutrient agar Peptone 5.0 g, beef extract 3.0 g, NaCl 5.0 g, agar 20 g, distilled water 1000 ml Trypticase soy agar (pH 7.3) Trypticase (animal peptone) 15 g, phytone (soy peptone) 5.0 g, NaCl 5.0 g, agar 20 g, distilled water 1000 ml Tryptone broth agar Tryptone 10.0 g, NaCl 5.0 g, $CaCl_2$* 1.0 g, agar 20 g, distilled water 1000 ml *Sterilized separately and mixed with medium just before pouring Simmons citrate agar (pH 6.9) Ammonium dihydrogen phosphate 1.0 g, dipotassium hydrogen phosphate 1.0, NaCl 5.0 g, sodium citrate 2.0 g, magnesium sulfate 0.2, bromothymol blue 0.08 g, agar 20 g, distilled water 1000 ml
2	Fungi	Starch agar (pH 7.0) Starch soluble 20.0 g, peptone 5.0 g, beef extract 3.0 g, agar 20 g, distilled water 1000 ml Sabouraud's dextrose agar (pH 5.6) Peptone 10.0 g, dextrose 40.0 g, agar 20 g, distilled water 1000 ml Potato dextrose agar (pH 7.3) Potato (peeled, cut)# 200.0 g, dextrose 20.0gm, agar 20 g, distilled water 1000 ml #Boil peeled, cut potatoes (200 g) in 500 ml water for 10 minutes, filter extract through muslin cloth, add dextrose and agar to filtrate, and make up the volume to 1000 ml Emmon's culture medium for fungi Glucose 20gm, neopeptone 10.0 g, agar 20 g, chloramphenicol^ 40.0 mg, distilled water 1000 ml ^ All other ingredients except chloramphenicol are boiled separately Chloramphenicol solution is prepared by suspending in10ml of 95% alcohol and then mixed to boiled ingredient and autoclave media for 10 minutes only. Media support growth of all types of mold and fungi
3	Actinomycetes	International *Streptomyces* Protocol medium-2 (ISP-2), dextrose 4.0 g, yeast extract 4.0 g, malt extract 10.0 g, agar 20.0 g, distilled water 1000 ml Czapek-Dox (modified) agar (pH 6.4–7.0) Sodium nitrate 2.0 g, KCl 0.5 g, magnesium sulfate 0.5, dipotassium hydrogen phosphate 1.0, ferrous sulfate 0.01 g, starch 10.0gm, peptone 10 g, agar 20 g, distilled water 1000 ml

[a]To prepare broth medium agar is not added in the medium composition

3. For example, to prepare 300 ml of nutrient agar, the amount of peptone required will be calculated as
   Peptone concentration for 1000 ml = 5.0 g
   Hence, for $\frac{300\ \text{ml}=5.0\ \text{g}\times 300\ \text{ml}=1.5\ \text{g}}{1000\ \text{ml}}$
4. Remove plug from mouth of flask and keep it aside.
5. Weigh media ingredients using weighing balance; firstly tear the weight of paper on which weighing is done; the sequence of weighing should be from nonhygroscopic (non-sticky) ingredient first to hygroscopic ingredients.
6. To weigh nutrient agar medium, weigh NaCl first followed by agar, peptone, and beef extract. While adding ingredient to flask, care should be taken to avoid sticking of medium components to walls. The water should be poured at last from walls of flask so as to dissolve any sticky medium ingredients.
7. Cover the mouth of flask with cotton plug and cover it with newspaper.
8. Put water in autoclave to dip the heating rod. Then place stand and steel basket over it. Keep the media flask in it.
9. Close tightly the autoclave lid and switch it on.
10. Autoclave media at 121 °C, 15 lbs., 15 min.
11. Switch off the autoclave and allow the pressure to be released completely, in natural way (pressure gauge reading = 0).
12. Unscrew the autoclave lid and take out the sterilized medium and transfer it to laminar airflow.
13. Place the sterilized Petri plate bundle in laminar airflow. Close the lid of laminar airflow. Switch on the UV light for 10–15 min.
14. Allow the media and Petri plate bundle to cool. When the flask becomes easy to hold in hand (45 °C), the media can be poured.
15. Switch off UV light and switch on fluorescent light; on blower, clean the surface of laminar airflow.
16. Pouring.
17. Keep Petri plates to left-hand side of burner and media flask to right-hand side.
18. Pick up the plate with left hand and flame the periphery quickly; do not overheat.
19. Open the plate at 45° angle facing toward flame.
20. Open the plug of media flask with the help of little finger and ring finger, hold the flask from right hand, and flame its neck.
21. Pour 25 ml of medium in each Petri plate and keep it on floor of laminar airflow.
22. Keep the poured plates in bundle of 5–8 plates, and at the top keep empty plate so that moisture does not get deposited on upper lid of media plates.

23. Allow the plates to solidify at room temperature.
24. Observe the plate after 24 hours for absence of contamination.
25. Use plates for culturing purpose.

Note:

1. Preparation and calculation of media for broth culture is same as solid medium except that the media is prepared in culture tubes and/ or conical flask. Also, the media used do not contain agar.
2. Heat-sensitive ingredients such as sugars are sterilized by filtration through 0.22 μm nitrocellulose filter and then mixed with other sterilized medium ingredients just before pouring.

Precautions:

1. Use separate papers for weighing different ingredients.
2. Always check and set level of watermark of weighing machine, before and after weighing.
3. Use separate spatula for weighing ingredients to avoid unnecessary mixing of ingredient.
4. Tighten the caps of boxes and bottles, to prevent ingredients from absorbing moisture and contamination.

Viva Voce:

Q1. Define (a) sterilization, (b) asepsis, (c) culture media, (d) inoculation, and (e) incubation.

Q2. Which method is most suited for sterilization of glassware?

Q3. Describe technique used for sterilization of media.

Q4. What are the basic ingredients of media used for growing microorganism?

Chapter 29

Sub-culturing of Bacteria, Fungi and Actinomycetes

Microbial growth is defined as increase in number and/or biomass. All microorganisms require food, oxygen, moisture, and space for growth. Like any other living organisms, they age, and their growth is inhibited in the absence of food, space, oxygen, etc. In order to sustain any microbial culture in laboratory conditions, sub-culturing is required. Sub-culturing is a procedure of transferring of microorganism into fresh nutritive medium from its stock culture. It includes transfer of culture from slant to slant, slant to plate, plate to plate, plate to slant, solid medium to broth, and broth to solid media. Sub-culturing is done to maintain culture in its active form (prolonging life and/or increase the number of cells) for varied applications.

Materials: Petri plates, filter paper, medium slant or plate containing pure culture of microorganism, Bunsen burner, 70% ethanol, sterilized cotton, Sabouraud dextrose agar (SDA) or any other medium, inoculation loop (for culturing of bacteria and actinomycetes) and inoculation needle (for culturing of fungi), fresh medium plates/slants/broth, laminar airflow.

Procedure:

1. Clean the surface of laminar airflow with 70% ethanol, close hood, and switch on the UV light.
2. Switch off the UV light after 10 min and open fluorescent light.
3. Switch on blower and open the hood and wipe the working surface with 70% ethanol.
4. Allow the surface to get dry for 2–3 min and ignite the Bunsen burner.
5. Dip the wire of inoculation loop (or needle) in 70% ethanol.

Aakanchha Jain et al. *Basic Techniques in Biochemistry, Microbiology and Molecular Biology: Principles and Techniques*, Springer Protocols Handbooks, https://doi.org/10.1007/978-1-4939-9861-6_29,

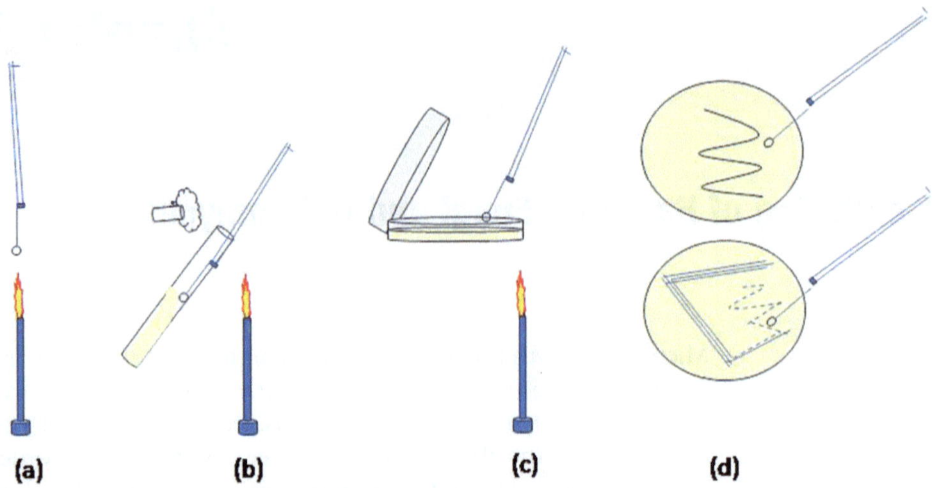

Fig. 1 Steps involved in sub-culturing of microorganisms (a) Sterilize the inoculation loop in upright position above the flame until it becomes red hot, then cool the loop in air; (b) Open the cotton plug of tube containing stock culture and flame the neck of tube with rotation, then insert the loop in slant and pick-up the culture of microorganism (bacteria, actinomycetes), close the mouth of tube with plug; (c) Open the media plate near flame and streak culture in any design like simple streak method, four streakmethod etc. Similar procedure is followed for sub-culturing of fungi but instead of inoculation loop, inoculation needle is used to pick culture and inspite of streaking, point inoculation is done at centre of plate

6. Take out the inoculation loop (or needle) from ethanol, and put it over the flame of the Bunsen burner in a manner as shown in Fig. 1a until it becomes red hot.
7. Cool the loop in air for 30 s, and remove the cotton plug of slant containing pure culture of microorganism.
8. Heat the neck of tube for 30 s.
9. Pick up the culture of microorganism with the help of inoculation loop (or needle), and inoculate it on media plates.
10. Close the plug of culture tube and open the upper lid of Petri plate at an angle of 45°.
11. Point inoculate fungi and streak bacteria on plates (Fig. 1).
12. Flame the inoculation loop/needle until it becomes red hot and place in test tube containing 70% (v/v) ethanol.
13. Incubate the inoculated plate/slant/broth in appropriate conditions:
    (a) Bacteria—37 °C for 24 h
    (b) Yeast—28 °C for 24–48 h
    (c) Fungi—28 °C for 2–4 days
    (d) Actinomycetes—28 °C for 2–4 days
14. Observations: growth of microorganism is visible on inoculated plates/ slant/ broth.

Precautions:

The loop and needle should be cooled in air properly else the culture may die.

Terminate culture in log phase of growth.

After flame sterilization of inoculation, loop/tube should never be kept at surface of laminar airflow.

The cotton plugs should be held in hand such that portion to be inserted inside tube mouth should be exposed in air, and the outer portion of plug should be held in hand; also never keep the plug on surface while working to prevent contamination (Fig. 1).

The culture plate and slant should be opened temporarily while transferring the culture.

Viva Voce:

Q1. Define microbial culture.

Q2. What is sub-culturing of microorganism?

Q3. Why should we sub-culture microorganism?

Q4. Define pure culture.

Q5. Sub-culturing of fungi is done with the help of loop/needle

Chapter 30

Preservation of Microorganisms: Stabs, Slants, Lyophilization and Cryopreservation

Microorganisms are continuously used for production of antibiotics, enzymes, vitamins, etc., biotransformation of biomolecules, bioremediation, bioconversion, bioprocessing of goods, etc. Thus, proper maintenance and preservation of microbial cells possessing desired potential (importance) is required to ensure long-term availability of them without loss in their metabolic potentials. The basic principal behind preservation of microbial cultures is to reduce their metabolism without disturbing cell integrity. Agar slant culture, agar slant culture covered with oil, stab culture, saline suspension, glycerol and DMSO preservation, drying on silica gel, drying on soil, sterile water, lyophilization, cryopreservation, etc. are methods employed for preservation of microorganism. However, the choice of method to be used depends on type of microorganism, purpose of storage, and duration of preservation.

Stab Culture: This method uses soft agar (10–12% agar) and contains only butt position. The culture is inoculated deep in medium with the help of sterilized inoculation needle. Use of stabs supports growth of microorganism (bacteria) and prevents drying of culture (Photograph 1).

Materials: Culture tubes, cotton plugs, Bunsen burner, cotton, ethanol (70%v/v), yeast extract malt extract medium ingredients, nutrient agar medium ingredients, autoclave, inoculation loop, inoculation needle, refrigerator, glass funnel, conical flask, hot plate, weighing machine, blotting paper.

Procedure:

1. Weight the medium ingredients in clean glass conical flask, and boil the medium ingredient for 10 min in hot plate. The soft

Aakanchha Jain et al. *Basic Techniques in Biochemistry, Microbiology and Molecular Biology: Principles and Techniques*, Springer Protocols Handbooks, https://doi.org/10.1007/978-1-4939-9861-6_30,

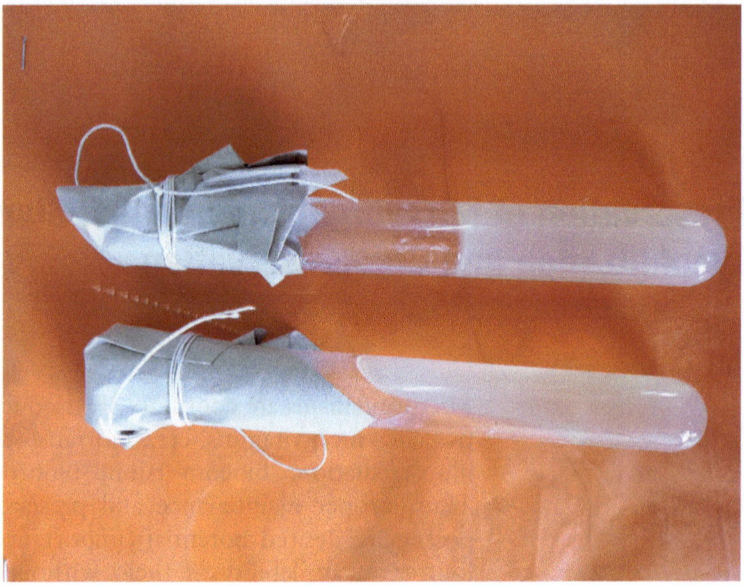

Photograph 1 Nutrient agar slant and stab

agar, i.e., 10–12% of agar, is supplemented in medium for preparation of stabs.

2. When temperature of medium cools down to 50 °C, pour it in culture tubes with the help of glass funnel.
3. Fill approximately 1/3 of tube volume with medium, and plug the tubes.
4. Make a bundle of seven tubes, cover it with paper, and autoclave the bundle of tubes in upright position at 121 °C for 15 min at 15 lbs.
5. After autoclaving allow the media in tubes to solidify.
6. Inoculate the stab with pure culture of microorganism using sterilized straight needle.
7. Incubate the tube at 37 °C for 24 h for bacterial culture, and refrigerate it at 5–10 °C.
8. The stab culture storage period in refrigerator is 2–3 months. Hence, sub-culturing should be done at regular interval of 2 months.

Agar Slant Culture: The agar slant can be divided into butt and slant/inclined portion where slant acts as a reservoir of nutrients and slant provides comparatively larger area for streaking of microorganism. On refrigeration the metabolic activities of cells reduce due to low temperature, but the activities do not get arrested, i.e., microorganisms keep on growing in refrigerated conditions but rate slows down. This method is one of the most common methods used for preservation of culture in teaching and research laboratories. It is important to note that media used for preservation should support growth of microorganism but should not allow the rapid growth.

Materials: Culture tubes, cotton plugs, Bunsen burner, cotton, ethanol (70%v/v), yeast extract malt extract medium ingredients, nutrient agar medium ingredients, autoclave, inoculation loop, inoculation needle, refrigerator, glass funnel, conical flask, hot plate, weighing machine, blotting paper.

Procedure:

1. Weight the medium ingredients in clean glass conical flask, and boil the medium ingredient for 10 min in hot plate.
2. When temperature of medium cools down to 50 °C, pour it in culture tubes with the help of glass funnel.
3. Fill approximately 1/3 of tube volume with medium.
4. Insert cotton plug in culture tubes, and make a bundle of seven tubes.
5. Cover the bundle with paper, and autoclave the bundle of tubes in upright position at 121 °C for 15 min at 15 lbs.
6. After autoclaving take out the bundle of tubes, and keep it in an 30° angle.
7. Allow the media in tubes to solidify.
8. Streak or point inoculate the slants with pure culture of microorganism.
9. Incubate the tube at appropriate conditions as mentioned below:

   (a) Bacteria—37 °C for 24 h

   (b) Yeast—28 °C for 24–48 h

   (c) Fungi—28 °C for 2–4 days

   (d) Actinomycetes—28 °C for 2–4 days
10. The preservation is done in refrigerator at 5–10 °C.
11. The slant culture storage period in refrigerator is 2–3 months.
12. Thereafter sub-culturing should be done.

Precautions: The overheating of medium should be avoided. Overheating may cause loss in nutrient value of medium.

Dispensing of medium in culture tube should be done with the help of funnel to avoid sticking of media in tube walls. Media in tube walls may cause problem in handling of culture and can act as a source for contaminating microorganisms.

Minimum two sets of culture should be made for each culture: (a) stock culture, only to be used for sub-culturing, and (b) working culture, used for routine work.

Lyophilization or Freeze-Drying: This method is used for long-term storage of microorganism and is based on principle of sublimation. The water in its frozen form is removed from sample by sublimation. Removing of water slows down the enzymatic reactions and reactions involving free radicals, etc. Spore-forming microorganisms can be preserved using lyophilization. The process of freeze-drying can be divided into (a) preparation of microorganism; (b) suspension of microorganism in a suitable freezing medium (skim milk or sucrose); and (c) freeze-drying process.

Materials: Culture tubes, glass ampoules, strips of filter paper, Pasteur pipettes, sterilized cotton wool (non-adsorbent) plugs, sterilized distilled water, air glassblowing torch, glass cutter in support handle, suspending medium [(a). skimmed milk (10–20% w/v solution); (b). trypticase soy broth, 0.7 gm, sucrose, 10 gm, bovine serum albumin, 5.0 gm, distilled water, 100 ml (filter sterilize through 0.22 μm filter); (c). sucrose, 20 gm, bovine serum albumin, 10.0 gm, distilled water, 100 ml (filter sterilize through 0.22 μm filter)], Bunsen burner, cotton, ethanol (70%v/v), freezer, freeze-drying apparatus, inoculation loop, inoculation needle, refrigerator.

Note: Milk solution should be autoclaved at 116 for 20 min.

Use of skimmed milk as suspending medium is not recommended for bacteria and fungi that get inhibited by milk.

Procedure:

1. Cut small strip of filter papers, and label name of microorganism on it with pencil.
2. Sterilize clean glass ampoules in hot air oven at 180 °C for 1–2 h. Allow it to cool to room temperature.
3. Prepare suspension of microorganism in appropriate suspending medium (skim milk, sucrose, etc.).

4. Transfer 0.25–0.5 ml of microbial culture (cell density ~10^9/ml) in sterilized glass ampoules.
5. Keep glass ampoules in freezer (−50 °C/−70 °C) for an hour.
6. Allow temperature of freeze dryer shelf to reach −40 °C, keep the frozen glass ampoules in it, and switch on the vacuum pump.
7. Initiate primary drying at −40 °C, and continue for 4 h.
8. Within 20 min vacuum pressure reaches 200 m torr. Once the vacuum pressure is below 200 m torr, raise the temperature of chamber at the rate of 10° increment.
9. Hold the tube for half an hour at each increased temperature. When temperature rises to 20 °C, hold the tube for 1 h and release vacuum.
10. Keep the ampoules at −70 °C overnight.
11. Subsequently, dry samples for 2 h, and release vacuum.
12. With the help of sterile Pasteur pipette, push the cotton wool plug inside ampoule up to 1/3 length of ampoule.
13. Narrow down the glass of ampoule above the cotton wool plug through flaming with rotation.
14. Glass becomes malleable and thereby can be stretched to narrow down channel at both ends; the constriction should be approximately 2 cm long.
15. Attach constricted ampoules to port of freeze dryer and on the vacuum pump for secondary drying.
16. Dehydration period may range from 2 to 12 h depending upon machine used and type of microorganism.
17. Seal the ampoule with the help of gas-air glassblowing torch.
18. Check ampoules for proper sealing using spark tester (vacuum tester).
19. Perfect vacuum is indicated when inside of tube glows purple.
20. Ampoules are stored at 4 °C for longer.
21. Revival of culture: flame the neck of ampoule, put a scratch at midway position of cotton wool ampoule, and with pressure break the sealing.
22. Remove cotton plug with the help of sterile forceps.
23. Pour 0.2 ml of sterile broth inside ampoule, and allow rehydration for a few seconds.
24. Transfer the entire content of tube with the help of Pasteur pipette in fresh medium.
25. Purity can be confirmed by serially diluting the culture followed by streaking.
26. The pure colonies are visible on plates.

Advantage:

Handling of ampoule is easy and can be stored for longer duration.

Note: Every culture behaves in a different manner; thus optimization of protocol should be done before using the method especially for genetically modified cultures.

Preservation in Liquid Nitrogen: Preserving culture in ultra-low temperature, i.e., −196 °C, in liquid nitrogen is known as cryopreservation. Storage at this temperature suspends the metabolism of cells and thus is considered best in terms of viability and genetic stability. Microbial culture is preserved in small ampoules along with stabilizing agents such as glycerol (10%v/v) and dimethyl sulfoxide (DMSO 10% v/v). Use of stabilizing agent prevents formation of ice crystals.

Materials required

Suspension of microorganism, DMSO, glycerol, sterilized plastic screw cap vials.

Procedure:

1. Make suspension of microorganism (density 10^8 Cells/ml) in glycerol (10% v/v).
2. Check the viability of culture suspension.
3. Label screw cap vials with cryo-marker pen.
4. Dispense 1 ml of suspension in sterilized plastic tube (tube should withstand ultracold temperature).
5. Perform programmed cooling at the rate of 2 °C per minute, or else put the vial at −70 °C in freezer for 2 h.
6. Transfer the culture in liquid nitrogen storage system (cylinder).
7. Check the viability of microorganism by thawing at 37 °C.

Viva Voce:

Q1. What are the methods used for preservation of microorganism?

Q2. Sub-culturing of slants kept in fridge should be done after how much duration?

Q3. Density of spores required to preserve microorganisms by lyophilization is ....

Q4. What is cryopreservation?

Q5. What is lyophilizer?

Chapter 31

Staining Methods – Simple Staining, Negative Staining, Gram's Staining and Acid-Fast Staining

Staining is a technique where specimens are stained with specific dyes to create contrast between the specimen and its background. Several different types of staining procedure are known today which are used to study morphological features of microorganism and their parts (Table 1).

Simple staining

Simple staining uses only one staining reagent and is used to determine shape, dimensions, and arrangement of microbial cells. Differential staining uses more than one staining reagent (dye) to differentiate cellular structures.

Materials: *Streptococcus*, *Escherichia coli* culture, methylene blue reagent, glass slide, inoculation loop, Bunsen burner, blotting paper, pasture pipette, staining tray, binocular microscope.

Methodology:

1. Take 1–2 drops of sterilized water on clean glass slide.
2. Pick culture of bacteria with the help of inoculation loop under aseptic conditions, and make suspension with water.
3. Prepare smear of bacteria by spreading it over slide.
4. Air-dry the smear and heat fix.
5. Spread 2–4 drops of staining reagent over the smear for 2–5 min.
6. Pour off the extra stain and was slide in running water.

Aakanchha Jain et al. *Basic Techniques in Biochemistry, Microbiology and Molecular Biology: Principles and Techniques*, Springer Protocols Handbooks, https://doi.org/10.1007/978-1-4939-9861-6_31,

Table 1
Different types of staining with examples

S. no.	Staining procedure category	Example
1	Simple staining	
1A	Bacteria	Crystal violet staining of actinomycetes/ bacteria
2	Differential staining	
2A	Group differentiation	Gram's staining
		Acid-fast staining
		Mycoplasma staining
2B	Structure morphology	Flagella staining
		Capsule staining
		Spore staining
		Nuclear staining
		Cytoplasm staining
3	Fungi	Lactophenol cotton blue staining of fungi

7. Keep the slide in inclined position, and remove water droplets carefully using blotting paper.
8. View the slide under low (100×), high (400×), and oil immersion (1000) magnification of compound microscope.
9. Observation: Blue-colored cocci cell arranged in the form of chain and blue-colored rod-shape cells are visible under oil immersion microscope. Draw the diagram of it, and note down the observations.

Negative staining of bacteria

The background of specimen (bacteria) gets stained instead of bacteria in negative staining; therefore, contrast is created. The acidic stains such as nigrosine and Indian ink are used. It relies on the principle that bacterial cell walls possess negative charge and thus repel acidic dye and do not get stained. Its biggest benefit is that cell morphology remains intact and even capsulated bacteria can be observed.

Materials: *Staphylococcus aureus*, nigrosine stain, glass slide, inoculation loop, Bunsen burner, blotting paper, pasture pipette, staining tray, binocular microscope.

Methodology:

1. Prepare suspension of bacteria in sterile water blank using inoculation loop, under aseptic condition.
2. With the help of pasture pipette, put 1–2 drops of bacterial suspension in clean glass slide, add 1 drop of nigrosine stain, and mix well with the help of loop.
3. Spread the suspension with the help of second slide. Stretch the slides from one end to another.
4. Dry the slide.
5. Observation: Examine slides under oil immersion microscope. Cells appear transparent against blue background. Draw the diagram of it, and note down the observations.

Gram's staining of bacteria

It is a type of differential staining, developed by Dr. H C Gram. It differentiates between gram-positive and gram-negative bacteria based on composition of their cell wall. The cell wall of gram-negative bacteria has high lipid content and that of gram-positive bacteria lacks it and has high content of peptidoglycan. The bacterial smear is stained sequentially with four reagents, namely, crystal violet (primary stain), iodine solution (mordant), alcohol (destaining solution), and safranin (counter stain). Lipid present in the cell wall of gram-negative bacteria solubilizes in presence of alcohol, and pores are created leading to loss of CV-I (crystal violet-iodine complex), while cell wall of gram-positive bacteria gets dehydrated on washing with alcohol and thus retains CV-I complex and appears violet. The gram-negative bacteria on staining with safranin appear pink.

Materials: *Staphylococcus aureus*, *E. coli* culture crystal violet, Gram's iodine solution, 95% ethyl alcohol, safranin, glass slide, inoculation loop, Bunsen burner, blotting paper, pasture pipette, staining tray, binocular microscope.

Methodology:

1. Take clean glass slide and prepare smear of bacteria.
2. Air-dry the smear.
3. Hold the slide with forceps and heat fix smear.

4. Put 3–5 drops of crystal violet dye over the smear and uniformly spread it over the smear, stain for 1 min.
5. Wash the slide with distilled water.
6. Spread Gram's iodine over the smear for 1 min.
7. Wash smear with 95% ethanol using pasture pipette; continue washing until release of color stops.
8. Wash with distilled water.
9. Put safranin over smear for 1 min and again wash the slides.
10. View the slide under low (100×), high (400×), and oil immersion (1000) magnification of compound microscope.
11. Observation: Pink-colored, rod-shaped cells are gram-negative rod (*E. coli*), and those appearing violet cocci in clumps are gram-positive (*Staphylococcus*) as viewed under oil immersion microscope. Draw the diagram of it, and note down the observations.

Acid-fast staining

Acid-fast staining is used for identification of *Mycobacterium* species. Their cell wall contains high concentration of lipid, hence making it hydrophobic and impermeable to stain like Gram's stain. These bacteria can be stained with carbol fuchsin (Ziehl-Neelsen method).The stain binds to the mycolic acid present in the cell wall of *Mycobacterium*. Such bacteria that retain the carbol fuchsin (pink) dye even after washing with strong acid are classified as acid-fast bacteria (resistant to acid-alcohol). Non-acid-fast bacteria lose the carbol fuchsin stain, and on applying methylene blue, they are counterstained as blue cells.

Materials required: *Mycobacterium* culture, *Staphylococcus aureus*, clean glass slide, inoculation loop, Bunsen burner, blotting paper, pasture pipette, carbol fuchsin stain, staining tray, methylene blue dye (5 gm/liter), 3% acid-alcohol, binocular microscope.

Methodology:

1. Make smear of *Staphylococcus aureus* and *Mycobacterium* culture in separate slides as mentioned before.
2. Add 10–15 drops of carbol fuchsin over smear, and heat the smear for 3–5 min.
3. On cooling, wash slides with distilled water.
4. Wash the smear with acid-alcohol for 30 s.
5. Again, wash the slides with excess of distilled water.
6. Flood smear with methylene blue stain for 2–3 min.
7. Wash with distilled water, and dry slides in air.

Table 2
Results as viewed under microscope

Reagent	Non-acid-fast bacteria	Acid-fast bacteria
Carbol fuchsin with heat	Dark pink	Dark pink
Acid-alcohol	Colorless	Pink
Methylene blue	Blue	Pink

8. View slide under low (100×), high (400×), and oil immersion (1000) magnification of compound microscope (Table 2).

Lactophenol Cotton Blue staining for fungi

Lactophenol cotton blue is used to stain fungal cultures: it contains three components basically (a) phenol that kills the fungi, (b) lactic acid that acts like a preservative, and (c) cotton blue that binds to fungal chitin and cellulose present in the cell wall of fungi and stains them blue.

Materials: Clean glass slide, inoculation needle, blotting paper, pasture pipette, lactophenol cotton blue dye.

Methodology:

1. Clean slide with 70% ethanol. Dry it.
2. Take one drop of lactophenol cotton blue dye on slide with the help of pasture pipette.
3. Pick the culture of fungi with the help of needle and forceps both from periphery of colony and from center; do not pick excess of culture.
4. Spread the culture in drop of lactophenol cotton blue.
5. Place the needle at 30 degree to surface of slide.
6. Pick up the coverslip with the help of forceps from the other hand, and put it over the needle.
7. Slowly move the needle out of stain such that the coverslip touches the dye and no air bubbles form.
8. Examine slide under the microscope (100× and 400×), note down the morphological characteristics, and draw diagram.
9. Observations: fungal mycelium and spore stain blue.

Viva Voce:

Q1. What is the need for staining of microorganisms?

Q2. Name two dyes used for simple staining.

Q3. Name two gram-positive bacteria and how they appear (color and shape) after Gram's staining.

Q4. Which stain is used to stain acid-fast bacteria?

Q5. What is smear fixing? And how is it done?

Chapter 32

Isolation of Coprophilous Fungi (Moist Chamber Method)

Dung of herbivores animal contains high content of organic matter and is rich source of metabolically diverse saprophytic microorganisms which hold high potentials for varied applications. Microbiologically dung consists of several microhabitat and thus support growth of different types of microorganisms. Some of the coprophilous (dung loving) fungi are *Poroniaerici, Cheilymenia* sp., *Cyathusstercoreus, Coprinus* sp., *Panaeolus* sp., *Psilocybe* sp.,*Pilobolus* sp. etc. They are generally slow growers and show visible growth in 10–15 days. Isolation of coprophilous fungi and studies related to their succession can be done using moist chamber method. Dung being rich source of nutrient act as medium for growth under moist conditions. They are of great importance for ecological research, biodiversity studies and for production of various bioactive compounds such as novel antibiotics, growth stimulators, biopesticides, enzyme.

Materials: Sterilized blotting paper, BOD incubator, Sterilized Petri plates, fine sterile needle, fresh dung of cow or horse, Sabouraud dextrose agar (SDA) or Potato dextrose agar media plates, Stereomicroscope.

Procedure:

1. Collect fresh dung sample and bring to laboratory.
2. Clean the working chamber (laminar air flow) with 70% spirit and lit the Bunsen burner.
3. Take sterilized Petri plate coated with blotting paper. Moisten the blotting paper with sterilized water. Perform the experiment in triplicate in two sets (total plates 6).
4. Transfer 10–15 g of fresh dung in beaker with the help of spatula. Level the surface of dung with help of spatula.
5. Cover First set (3 plates) with black paper from outside.

Aakanchha Jain et al. *Basic Techniques in Biochemistry, Microbiology and Molecular Biology: Principles and Techniques*, Springer Protocols Handbooks, https://doi.org/10.1007/978-1-4939-9861-6_32,

6. Incubate all the plates in BOD incubators at 25–30 °C for 10–15 days.
7. Observations:
8. The plates are examined under stereomicroscope for growth of fungal hyphae and sporulation. The plates covered with black paper when exposed to light may show spore dispersal of certain fungi like *Pilobolus* species.
9. The mycelium and spore are picked with sterilized inoculation needle and transferred to SDA or PDA plates.

Precautions:

1. The moisture should be maintained throughout the isolation process.
2. The media used for purification of fungi should be supplemented with antibacterial antibiotic to avoid bacterial contamination.

Viva Voce:

Q1. What is dung fungi.

Q2. What is growth period of dung fungi.

Q3. Name some coprophilous fungi.

Q4 Applications of coprophilous fungi.

Q5 Name technique used for isolation of coprophilous fungi.

Chapter 33

Isolation of Microorganisms from Air

Unlike soil and water, air does not contain nutrient material to support growth of microorganism; it acts as medium for transfer of microbial propagules. Microbes in air generally come from soil, water, and living organisms' activities. The microflora of air is highly dynamic and is affected by temperature, wind speed, moisture/humidity, pollution, and other human and animal activities. They are of great importance as their presence in air may exert adverse effects to the health of population and may cause spoilage, promote composting, biodegradation, etc. The qualitative and quantitative estimation of microorganism present in air is possible using sampling and isolation technique. In exposure plate technique, media plates are exposed in air for specified duration, and the microbial flora settles down on plate. When plates are incubated, the colonies of microorganism develop on plate, which can be further purified and identified.

Materials: PDA media plates, stop clock, incubator, glass marker, 70% ethanol, colony counter.

Procedure:

1. Mark sterilized media plates according to period of aerial exposure time.
2. Potato dextrose agar medium is recommended for isolation of fungi, nutrient agar medium for isolation of bacteria, and Thornton medium for isolation of actinomycetes.
3. Expose plates in air by removing upper lid of Petri plate at a height of human breathing zone for different duration, e.g., 1, 2, 3, 4, and 5 min.
4. Cover lower lid with upper lid and bring the plates to laboratory, and incubate them for desired period at specific temperature.

Aakanchha Jain et al. *Basic Techniques in Biochemistry, Microbiology and Molecular Biology: Principles and Techniques*, Springer Protocols Handbooks, https://doi.org/10.1007/978-1-4939-9861-6_33,

Table 1
Observation Table

	Number of colonies			
Exposure time (minutes)	Set 1a	Set 1b	Set 1c	Set 2a
1				
2				
3				
4				

5. Experiment is performed triplicate.
6. Observe plates for counting of colonies and their types. Record the results in tabular form (Table 1).

Viva Voce:

Q1. Name microorganisms found in outdoor and indoor environment.

Q2. Importance of microorganism in indoor environment.

Q3. Importance of microorganism in outdoor environment.

Q4. Which technique is used for isolation of microorganism from air?

Chapter 34

Motility Testing – Hanging Drop Method and Stab

The mobility in bacteria is largely credited to presence of large hair-like appendages over the body called as "flagella." These are the locomotor organ of bacteria. The process of staining the flagella is very tedious. Also it is very difficult to observe in the normal compound microscope. Bacterial movement must be distinguishable from the "Brownian movement" (the movement of small invisible particles in any medium). However, the presences of bacteria/activity of flagella can be visualized by the ability of microbe to move in growth medium or by hanging drop technique.

Materials: A broth culture of motile bacteria, *Proteus vulgaris*, approx. 18 h old; cavity slides, coverslip, petroleum jelly, soft agar medium.

Procedure:

(a) Hanging drop technique.

1. Apply a thin layer of petroleum jelly around the cavity circumference of slide or alternatively; apply petroleum jelly on two alternative side of cover slip as shown in Photograph 1.
2. Place a very small drop of culture on the cover slip.
3. Invert the concave side of slide over the cover slip without disturbing the culture drop.
4. Quickly turn the entire preparation upside down so that the suspension drop hangs in the cavity.
5. Bring the corner of the hanging drop in low power, i.e., 10× focus. Once the focus is set, turn to high-power objective.

Aakanchha Jain et al. *Basic Techniques in Biochemistry, Microbiology and Molecular Biology: Principles and Techniques*, Springer Protocols Handbooks, https://doi.org/10.1007/978-1-4939-9861-6_34,

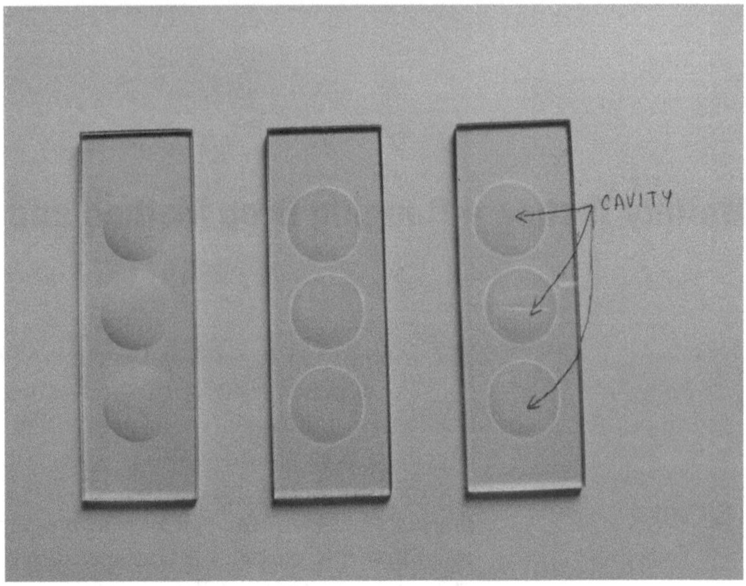

Photograph 1 Cavity slide for hanging drop method

(b) Stab culture technique.

1. Prepare 0.7% w/v nutrient agar medium; sterilize at 121 °C for 15 min at 15 lbs.
2. Pour the medium in aseptic conditions in test tubes (stabs).
3. After 24 h, inoculate the broth with loopful of culture in laminar air flow.
4. Incubate the tubes for 48 h at 37 °C.

Result: If the medium appears cloudy after incubation, the bacteria inoculated are motile. Non-motile bacteria grow at the surface or line of stab.

Viva Voce:

Q1. Which organelle is responsible for motility in bacteria?

Q2. Name the techniques used for motility determination.

Q3. What is the purpose of hanging the drop for motility determination?

Q4. What is the major difference between slide used for hanging drop technique and simple slide?

Q5. Name any two bacteria which are motile.

Chapter 35

Sterility Testing of Pharmaceuticals

Principle: The main concept of test for sterility is based on the fact that if microbes are inoculated in a clear medium which supports their growth (i.e., optimum temperature, pH, nutrition, and water), the presence of microbes can be detected by change in turbidity of medium, provided that the test is performed in uncontaminated conditions. The interpretation of test is based on the fact that each container in a batch is the same.

Culture medium used for test:

(A) Fluid thioglycollate medium at 25 °C; pH 7.1 ± 0.2 [intended for the examination of clear liquid or water-soluble materials and to check the sterility of stored blood in blood banks] (Table 1).

Incubation conditions: 30–35 °C, aerobically.

(B) Alternative thioglycollate medium at 25 °C; pH 7.1 ± 0.2 [for sterility testing of turbid or viscous biological products].

Incubation conditions: 30–35 °C, anaerobically (Table 2).

(C) Soyabean-casein digest medium at 25 °C; pH 7.3 ± 0.2 [for sterility testing of molds and lower bacteria].

Incubation condition: 20–25 °C aerobically (Table 3).

The media used for the test must comply with the following test during or after the sampling is being done:

1. Sterility: All the above three mentioned media should show no growth of microbes when incubated for 7 days.
2. Growth promotion test: For this test all the three media are inoculated with desired number (approx. 100) of respective microbes as given in Tables 1, 2 and 3 in duplicate. Clear growth should be visible after 7 days of incubation.
3. Test for bacteriostasis and fungistasis:

   (i) Take the microbial culture and inoculate about 100 viable organisms in respective medium given in Tables 4 and 5

Aakanchha Jain et al. *Basic Techniques in Biochemistry, Microbiology and Molecular Biology: Principles and Techniques*, Springer Protocols Handbooks, https://doi.org/10.1007/978-1-4939-9861-6_35,

Table 1
Fluid thioglycollate medium composition

Ingredients	Gms/liter	Purpose
Agar (moisture <15%)	0.75	Favors the growth of aerobes as well as anaerobes
Dextrose (glucose)	5.5	Growth for bacterial multiplication
L-cystine	0.5	Lower the oxidation-reduction potential; growth for bacterial multiplication
Resazurin sodium (0.10%)	1.0 ml	Color indicator
Sodium chloride	2.5	Osmotic equilibrium
Sodium thioglycollate	0.5	Reducing agent and neutralizes the toxic effects of mercurial
Tryptone	15.0	Source of nitrogen and carbon
Water-soluble yeast extract	5.0	Growth for bacterial multiplication

Table 2
Alternative fluid thioglycollate medium composition

Ingredients	Gms/liter	Purpose
Dextrose (glucose)	5.5	Growth for bacterial multiplication
L-cystine	0.5	Lower the oxidation-reduction potential; growth for bacterial multiplication
Pancreatic digest of casein	15.0	Source of amino acids and peptides
Sodium chloride	2.5	Osmotic equilibrium
Sodium thioglycollate	0.5	Reducing agent and neutralizes the toxic effects of mercurial
Tryptone	15.0	Source of nitrogen and carbon
Water-soluble yeast extract	5.0	Growth for bacterial multiplication

Table 3
Soyabean-casein digest medium composition

Ingredients	Gms/liter	Purpose
Dextrose	2.5	Growth for bacterial multiplication
Dipotassium hydrogen phosphate	2.5	Lower the oxidation-reduction potential; growth for bacterial multiplication
Pancreatic digest of casein	17.0	Source of amino acids and peptides
Papaic digest of soyabean meal	3.0	Reducing agent and neutralizes the toxic effects of mercurial
Sodium chloride	5.0	Osmotic equilibrium

Table 4
Microbes and their corresponding growth medium used for growth promotion test and test for bacteriostasis and fungistasis

Test Microbes	Medium used[a]
Bacillus subtilis (ATCC 6633)	Fluid thioglycollate; soyabean-casein digest
Candida albicans (ATCC 10231)	Fluid thioglycollate; soyabean-casein digest
Bacteroides vulgates (ATCC 8482)	Fluid thioglycollate; alternative thioglycollate

[a]The incubation conditions are mentioned above

(ii) Add specified portion of preparation given in Table 5 being examined to half of the plates.

(iii) Incubate the plates at appropriate temperature for not more than 7 days and observe the growth (Tables 4 and 5).

Method A: Membrane Filtration Test: *Used for oil; ointment that can be put into solution; a non-bacteriostatic solid not readily soluble in medium used for test; a soluble powder or a liquid that have inherent bacteriostatic and fungistatic effect (Table 6).*

Apparatus and procedure of membrane filtration is depicted in Fig. 1.

Table 5
Sample volume test for bacteriostasis and fungistasis

Containers content	Quantity to be used	Culture medium	Method used
For liquids			
Less than 1 ml	Entire contents	15 ml	
1–5 ml	Half of the volume	20 ml	
5–20 ml	2 ml		
20–50 ml	5 ml	40 ml	
50–100 ml	10 ml or 10% of contents	80 ml	
More than 100 ml	Not less than 50% of contents		Method A
For solids			
Less than 50 mg	Entire contents	40 ml	
50–200 mg	Half of the contents	80 ml	
200 mg or more	100 mg		

Table 6
Diluting fluids for method A

	Quantity [pH 7.1 ± 0.2]	
Ingredients	Fluid A	Fluid B
Peptic digest of animal tissue	1.0 gm	1.0 gm
Sterile distilled water (q.s.)	1000 ml	1000 ml
Polysorbate 80	–	1.0 ml

1 Method B: Direct Inoculation (Table 7)

Interpretation:

- If no growth or traces are seen, the preparation passes the test.
- If growth is seen, preserve the container for further verification by other means. After verification a retest is done in same conditions but with different samples.
- If in the second test, no growth is seen, the preparation passes the test. However, if growth is seen and the growing microbe is same as that seen in the first test, the preparation fails the test.

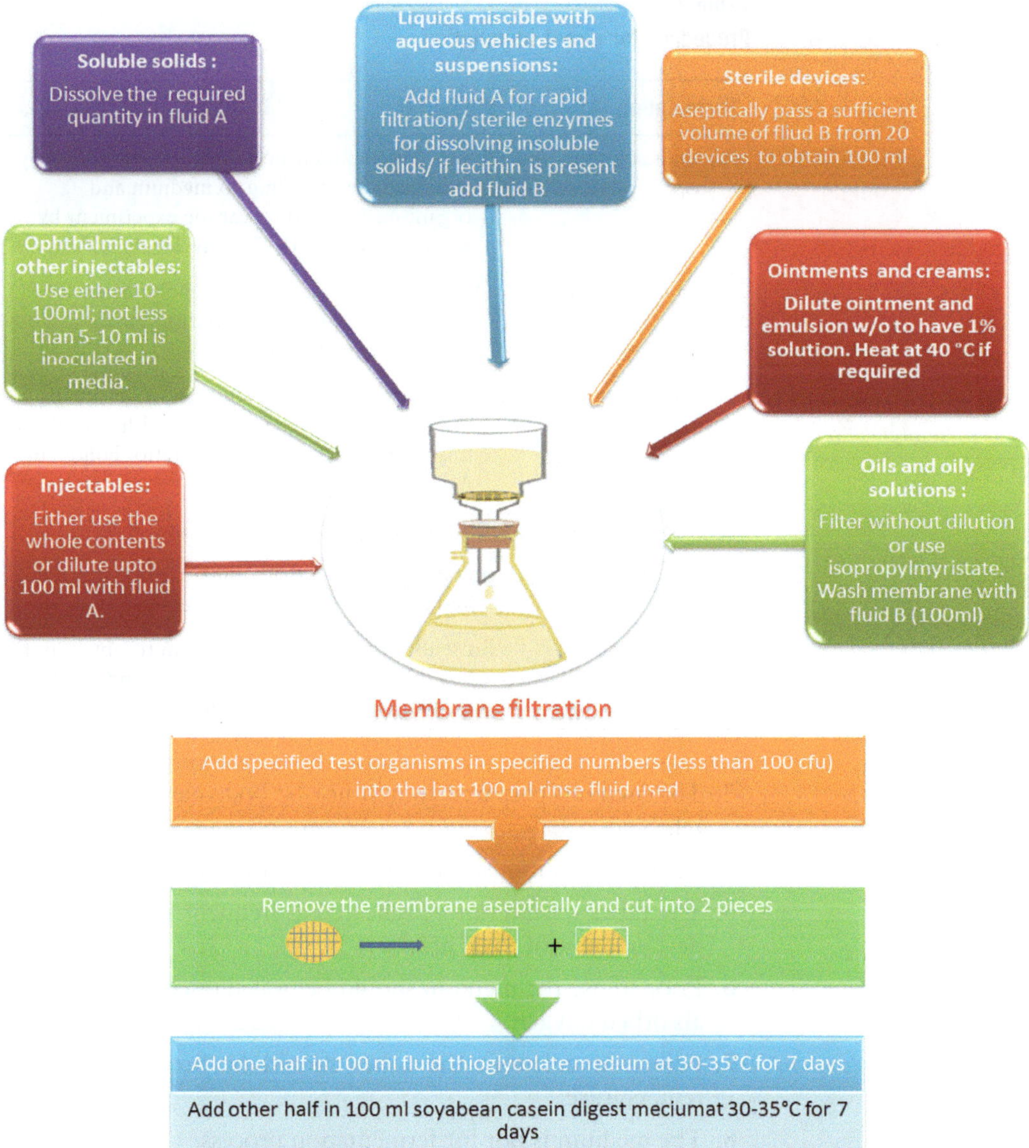

Fig. 1 Flow diagram of process of membrane filtration

- If in the second test, no growth is seen, the preparation passes the test. However, if growth is seen and the growing microbe is different as that seen in the first test, the test is repeated with twice number of samples.

Precautions:

1. The test should not adversely affect the growth of microbe used for testing of samples.
2. The air and surface of the working desk should be monitored regularly.

Table 7
Procedure for direct inoculation

Formulation	Method B
Aqueous solutions and suspensions	Remove the liquid from test container aseptically and incubate according to medium and organism. If turbid repeat the experiment by sub-culturing at 4th–7th day, for next 7 days
Oils and oily solutions	Media is added with 0.1% w/v polysorbate 80 or (4-tert-octylphenoxy) polyethoxyethanol. Follow the method as above
Ointments	Prepare ten-time diluted solution of ointments insoluble in isopropyl myristate with fluid B and mix in culture medium in 1:8 ratio. Follow the method as above
Soluble solids (both)	Transfer the quantity as given in table above
Sterile devices	Immerse in not less than 1000 ml of culture media and test further. Follow the method as above
Transfusion and infusion assemblies	Rinse the entire unit with medium to obtain not less than 15 ml (from 20 devices) and incubate in 100 ml. Follow the method as above.

3. The medium prepared should not be boiled again but can be reheated.
4. Resazurin sodium should be added at the end.
5. The tubes should not be reheated as frequent boiling leads to development of toxic products.
6. Prior to use the medium should be boiled once to remove the absorbed oxygen.
7. Before inoculation, the tubes should be brought to room temperature.
8. The medium is not for fermentation process.

Viva Voce:

Q1. Name the medium used for sterility testing of pharmaceuticals.

Q2. Elaborate the purpose of using the following in medium used for sterility testing of pharmaceuticals:

Resazurin sodium (0.10%).

Sodium chloride.

Sodium thioglycollate.

Dextrose.

Dipotassium hydrogen phosphate.

Pancreatic digest of casein.

Q3. Name the microbes and their corresponding growth medium used for growth promotion test.

Q4. Why should Resazurin be added at the end of media preparation?

Q5. State whether true or false.

(i) The medium prepared can be reboiled whenever needed.............

(ii) If in the second test, no growth is seen the preparation passes the test...............

(iii) If growth is seen in second test and the growing microbe is different as that seen in first test, the test is passed............

(iv) Sterile devices are tested by membrane filtration test.................

Q6. Give the composition of Fluid A in direct inoculation method.

Chapter 36

Physical, Chemical and Bacteriological Analysis of Water

(A) Physical and Chemical Analysis of Water

Constituents responsible for rendering water unfit for use includes clay, silt, humus, and color (aesthetically not acceptable and palatability decreases); pH (health-related problems); hardness (affect mucous membrane), TDS, Ca, Mg, and SO_4 (gastrointestinal irritation); fluoride (dental and skeletal fluorosis); nitrate, Ca, Mg, and Cl (methemoglobinemia and encrustation in water supply); and various analysis parameters of water are shown in Fig. 1.

Procedure: The physical measurement of water includes pH, specific gravity, specific conductance, acidity, alkalinity, hardness of water, nitrogen in water, total solids, and turbidity of water.

(a) **pH**: The pH is measured accurately using digital pH meter. The pH of pure water is between 4 and 9.

(b) **Specific gravity**: Specific gravity bottle is used for the determination.

1. Firstly, the bottle is filled up to the rim in a way so that when cap is placed excess fluid is removed. This does not allow air bubble entrapment. Note the weight.
2. Repeat the step with water to be tested and record the weight again. Apply the formula

$$\begin{aligned}&\text{Specific gravity of water from unknown source}\\&\quad = (\text{weight of sample})/\\&\qquad(\text{weight of standard} \times \text{specific gravity of water})\end{aligned} \tag{1}$$

The specific gravity may increase in case of contaminated water due to high ionic content.

Aakanchha Jain et al. *Basic Techniques in Biochemistry, Microbiology and Molecular Biology: Principles and Techniques*, Springer Protocols Handbooks, https://doi.org/10.1007/978-1-4939-9861-6_36,

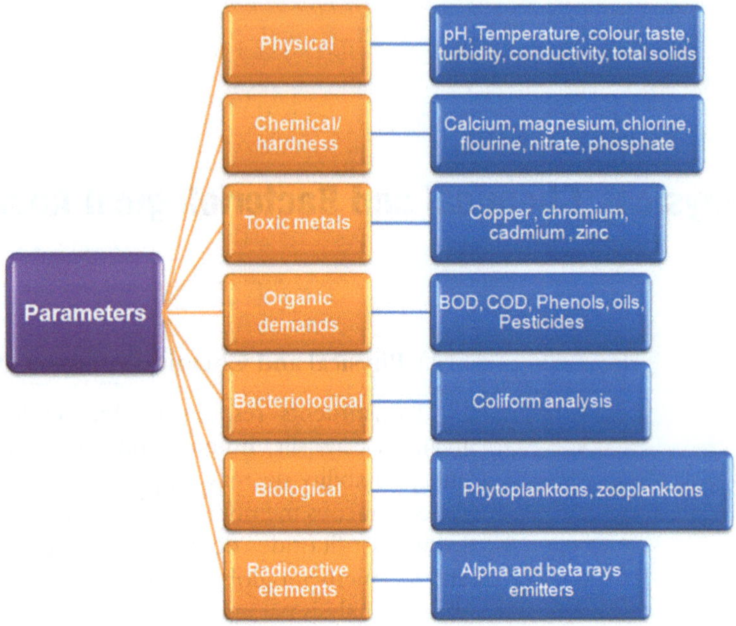

Fig. 1 Parameters of water analysis

(c) **Specific conductance**: The specific conductance is measured accurately using digital conductivity meter. The conductance may increase in case of contaminated water due to high ionic content.

(d) **Acidity**: It is measured titrimetrically using 0.1 N NaOH solution as titrant and phenolphthalein as indicator. Briefly, take 10 ml of sample water in conical flask. Add phenolphthalein and perform titration until pink color persists. Calculate the amount of NaOH used and compare it with standard water.

(e) **Alkalinity**: It is measured in the same way as acidity except 0.1 N HCl is used as titrant and methyl red/methyl orange as indicator.

Procedure:

1. Select the titrant.
2. Select the normality.
3. Adjust the speed of the drops from the burette.
4. Select the titrate and choose a definite volume of the water sample.
5. Select the indicator, phenolphthalein, to get a pink color.
6. Stop titration when the solution becomes colorless, and calculate phenolphthalein alkalinity (**PA**) as $CaCO_3$ (mg/L) using

the equation. Let **A** be the volume of titrant (mL) used in the titration (V1).

7. Add methyl orange to the same flask and continue titration till the color changes from yellow to orange. The total volume of titrant corresponds to total alkalinity (**TA**) as $CaCO_3$ (mg/L). Let **B** be the total volume of titrant (mL) consumed with both the indicators (V2) (Table 1).

Volume of HCl corresponding to phenolphthalein end point (A) = ...mL.

Normality of acid = ...N.

Volume of the water sample = ...mL.

Normality of water corresponding to phenolphthalein end point = ...ppm.

Volume of HCl corresponding to methyl orange end point (B) = ...mL.

Normality of acid = ...N.

Volume of the water sample = ...mL.

Normality of water corresponding to methyl orange end point = ...ppm (Table 2).

Results: Alkalinity is due to........... =.............ppm.

Table 1
Observation table for measuring alkalinity of water

		Burette reading (mL)		
No	Vol. of the sample (mL)	Initial	Final	Vol. of HCl (mL)
1			(A)	
2			(B)	

Table 2
Observation table for measuring alkalinity of water in terms of OH ions, CO_3, and HCO_3

Result of titration	OH alkalinity as $CaCO_3$	CO_3 alkalinity as $CaCO_3$	HCO_3 alkalinity as $CaCO_3$
PA = 0	0	0	TA
PA < 1/2TA	0	2PA	TA—2PA
PA = 1/2TA	0	2PA	0
PA > 1/2TA	2PA—TA	2(TA—PA)	0
PA = TA	TA	0	0

(f) **Hardness of water**

1. Select the titrant.
2. Adjust the speed of the drops from the burette.
3. Adjust the molarity of titrant.
4. Select a definite volume of water sample.
5. Choose the indicator and start the titration.
6. When color changes from wine red to blue, click the "stop" button and note the volume of EDTA used.
7. Then calculate the hardness of water sample in ppm using the equation as follows (Table 3):

$$\text{Total hardness as } CaCO_3(\text{ppm}) = \frac{\text{Volume of EDTA} \times 0.1 \times \text{molarity of EDTA} \times 10^6}{\text{Volume of the sample}} \quad (2)$$

Volume of EDTA used = ...mL.

Molarity of EDTA = ...M.

Volume of the water sample = ...mL.

Therefore the total hardness of the sample is = =................ppm.

Result: Total hardness of the water sample = ...ppm.

(g) **Nitrogen in water**

Sources like decayed vegetable, fertilizers and manure, animal feedlots, municipal wastewater, and sludge disposal to land leachates from refuse dumps, industrial discharges, septic systems, and n-fixation from atmosphere by bacteria and lightning which causes nitrogen contaminants in water.

Table 3
Observations table for measuring hardness of water

		Burette reading (mL)		
No	**Vol. of the sample (mL)**	**Initial**	**Final**	**Vol. of EDTA (mL)**
1				
2				

Methods

Ultraviolet Spectrophotometric Method

Useful for uncontaminated natural waters and potable water supplies that have low organic content. Follows Beer's law up to 11 mg/L as N. Dissolved organic matter, surfactants, nitrite, and hexavalent chromium may interfere with the test.

Procedure:

- Filter the sample.
- Add 1 ml of 1 N HCl per 50 ml of sample.
- Read absorbance or transmittance at 220 nm and 275 nm.
- Set 0 absorbance or 100% transmittance with distilled water.

Nitrate Electrode Method

Useful for nitrate concentration range of 0.14–1400 mg/L NO_3-N. Chloride and bicarbonate with weight ratios to NO_3-N > 10 or > 5, respectively, may interfere with the test.

Phenoldisulfonic Acid (PDA) Method

Nitrate reacts with phenoldisulfonic acid to produce nitro derivatives that in alkaline solution rearrange its structure to form yellow-colored compound with characteristics that follows Beer's law. Chloride interferes seriously which can be overcome by precipitation of chloride with Ag + as AgCl.

(h) **Total dissolved solids**: It is the portion of solids which pass through the 2.0 μm pore size filter or smaller. The determination involves two basic methods:

Evaporation at 180 ± 2 °C, at this temperature all the mechanically occluded water is lost.

Conductivity determination using digital conductivity meter.

(i) **Turbidity of water**

Turbidity in water is caused by suspended matter such as clay, silt, finely divided organic and inorganic matter, soluble colored organic compounds, plankton, and other microscopic organisms. Turbidity is an expression of the optical property that causes light to be scattered and absorbed. Turbidity measurement by nephelometer is written in terms of nephelometric turbidity units (NTU). Formazin polymer is used as the reference turbidity standard suspension.

Principle: Nephelometric method of turbidity measurement is based on a comparison of the intensity of light scattered by the sample under defined conditions with the intensity of light scattered by a standard reference suspension under the same conditions. The higher the intensity of scattered light, the higher the turbidity.

Precautions:

(a) Meter is designed to prohibit stray light reaching to detector.

(b) Short warm period is necessary to make the instrument free from significant drift.

(c) Clear colorless glass tube is used for sample.

(j) **Determination of chemical oxygen demand (COD) of water sample**

1. Select the water sample.
2. To reflux the contents in the RB flask, click the "switch on mantle" button.
3. Click "start titration" to titrate the contents.
4. Select the normality of ferrous ammonium sulfate (FAS).
5. Start titration and note the volume of titrant consumed when color changes from bluish green to wine red. (Let the volume of titrant be V2 mL.)
6. Repeat the same with the blank (let the volume of the titrant be V1 mL).
7. COD calculated using the equation (Table 4).

Volume of FAS used = (V1–V2) = ...mL.

Normality of FAS = ...N.

Volume of the water sample = ...mL.

Therefore COD of the water sample = ...ppm.

Result: COD of water sample = ...ppm.

Precautions:

1. Always wear lab coat and gloves in the lab.
2. Clean all the apparatus with chromic acid and distilled water and ensure that all the glasswares are free from water droplets while performing the experiment.

Table 4
Observation table for measuring COD

			Burette reading (mL)		
Sample	**No**	**Vol. of sample (mL)**	**Initial**	**Final**	**Vol. of FAS (mL)**
Sample	1	20	0		
	2	20	0		
Blank	1	20	0		
	2	20	0		

(B) **Bacteriological Analysis of Water**

Determination of the type of bacteria in water is more important than detecting its number. The three main contaminants mostly contributed by intestinal discharge are *E. coli*, *Streptococcus faecalis*, and *Clostridium* spp. *E. coli* being excreted in largest amount (200–400 billion per day) is of prime concern. The closely related forms are called as "coliforms." The membrane filtration test described in Experiment 3.8 may also be used for the same. The main method used for determination is called as "multiple tube fermentation test." Coliforms are aerobic or facultative anaerobic, gram-negative, rod-shaped, non-endospore-forming bacteria. It can ferment lactose in medium to produce acid and gas just in 24 h at 37 °C. The test is performed in three stages in sequence: presumptive, confirmed, and completed.

(I) Presumptive test

Principle: The acid and gas produced in lactose broth fermentation by coliforms are qualitated and separated by addition of bile, lauryl sulfate, and brilliant green in the medium. The result obtained is expressed as "MPN" (most probable number).

Materials:

- 9 ml lactose fermentation broth tube [LFB] (ten tubes).
- 20 ml double strength lactose fermentation broth tube [dsLFB] (three tubes).
- Sterile pipettes (10, 1, 0.1 ml).

Procedure:

1. Label three LFB tubes as "0.1," another three tubes as "1," and three tubes as ds LFB as "10."
2. Aseptically inoculate each "0.1", "1," and "10" tube with 0.1 ml, 1 ml, and 10 ml of water sample, respectively.
3. Incubate all the tubes at 37 °C for 24–48 h.
4. Observe the tube for production of acid (yellow color) and gas.

Results: If no gas is produced, presumptive test is negative; if gas is produced after 48 h, test is doubtful. The tubes showing positive test are kept for confirmation test.

(II) Confirmed test

Principle: In this test tubes positive or doubtful for presumptive test are streaked in selective differential medium for coliforms [eosin methylene blue (EMB)]. The dye in this medium is specific for growth of gram-negative bacteria but inhibits gram-positive bacteria. The test is positive if colored colonies are seen.

Materials: EMB agar plate; positive lactose broth culture from presumptive test.

Procedure:

1. Inoculate the EMB agar plate with culture from positive lactose broth.
2. Incubate the plate at 37 °C for 24–48 h in inverted positions.
3. Observe the tube for growth of *E. coli*.

Results: Appearance of colonies with dark center and metallic sheen indicates positive test.

(III) Complete coliform test

Lactose-positive colonies from EMB plate are isolated and subcultured. Gram staining is performed from these grown colonies.

Materials:

- 24 h, coliform positive, EMB agar culture from II test.
- Lactose fermentation broth.
- Nutrient agar slant.
- Gram staining reagents.

Procedure:

1. Inoculate the lactose fermentation broth with the isolated colony from EMB.
2. Streak nutrient agar slant with the isolated colonies from EMB agar plate.
3. Incubate both the broth and slant at 37 °C for 24–48 h.
4. Use the colonies in slant to perform gram staining as mentioned in section………….

Results: If the rod-shaped, gram-negative colonies are seen which produce gas and acid in lactose broth, the presence of coliform is confirmed.

Viva Voce:

Q1. The pH of pure water is………….

Q2. Hardness of water is mainly due to…………….

Q3. Define total dissolved solids.

Q4. ………………………..is used as standard in turbidity estimation of unknown samples.

Q5. Bacteriological analysis is mainly based on detection of …………………… bacteria.

Q6. State the principle of presumptive test.

Q7. ……………… .medium is used for confirmed coliform test.

Q8. ……………… .medium is used for complete confirmed coliform test.

Q9. State whether true or false.

(i) Dark centric colonies indicate positive confirmed test………

(ii) The incubation time for complete coliform test is 24–48 h……….

Chapter 37

To Perform Biochemical Identification of Microorganism: IMViC (Oxidative Fermentation and Carbohydrate Source Utilization)

Full Form: Indole, Methyl Red, Voges-Proskauer, Citrate (IMViC) tests.

Principle: The IMViC series of reactions allows for the differentiation of the various members of Enterobacteriaceae.

(A) Indole test
   (i) Certain microorganisms can metabolize tryptophan by tryptophanase.
   (ii) The enzymatic degradation leads to the formation of pyruvic acid, indole, and ammonia.
   (iii) The presence of indole is detected by addition of Kovacs reagent.

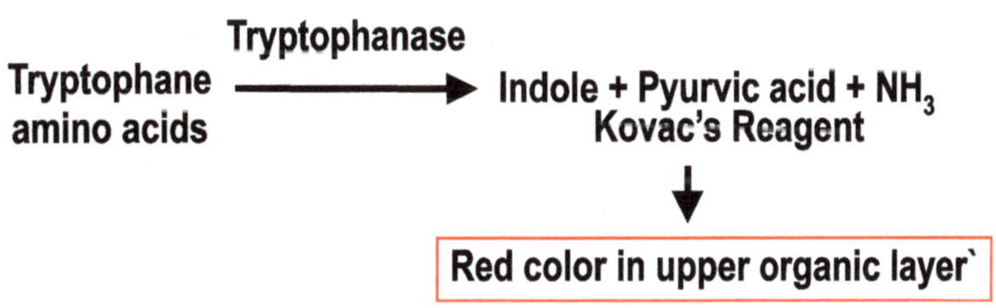

(B) I*MV*i*C* test or *M*ethyl *R*ed-*V*oges-*P*roskauer (MR-VP) tests are shown in Fig. 1.

Aakanchha Jain et al. *Basic Techniques in Biochemistry, Microbiology and Molecular Biology: Principles and Techniques*, Springer Protocols Handbooks, https://doi.org/10.1007/978-1-4939-9861-6_37,

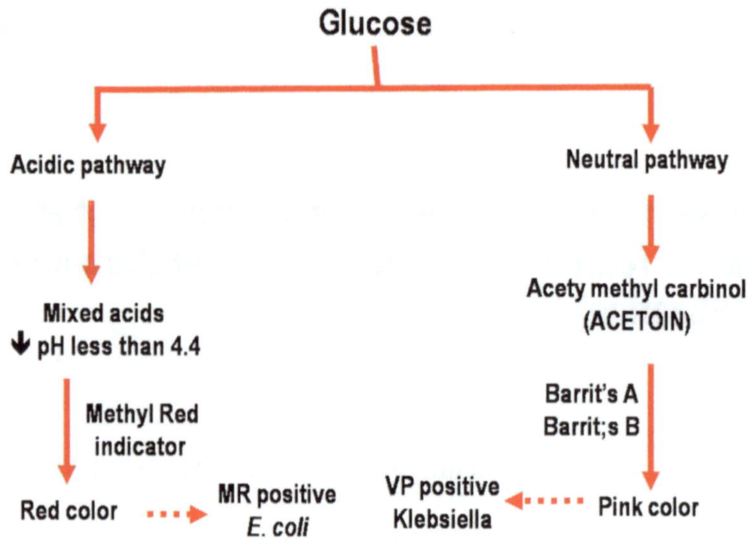

Fig. 1 MRVP test procedure

Procedure:

(A) Indole test

1. Inoculate tryptone water with the tested microorganism (works for *Escherichia coli*, *Enterobacter aerogenes*, *Proteus vulgaris*, and *Pseudomonas aeruginosa*).
2. Incubate at 37 °C for 24 h∗.
3. After incubation interval, add 1 ml Kovacs reagent; shake the tube gently and read immediately.

   ∗The reading should be done 48 h after inoculation, because over-incubation can allow the organism to utilize all the tryptophan and then degrade the indole, producing a false negative.

(B) I*MV*iC test or *M*ethyl *R*ed-*V*oges-*P*roskauer (MR-VP) tests

1. Inoculate the tested organism (*Escherichia coli*, *Enterobacter aerogenes*, *Proteus vulgaris*, *Pseudomonas aeruginosa*) into one tube of MRVP broth.
2. Incubate the tubes at 37 °C for 24 h.
3. After incubation: Pour 1/3 of the suspension into a clean nonsterile tube.
4. Run the MR test in the tube with 2/3 and the VP test in the open tube with 1/3.
5. For methyl red: Add 6–8 drops of methyl red reagent.

6. For Voges-Proskauer: Mix 12 drops of Barritt's A (α-naphthol) with 4 drops of Barritt's B (40% KOH); shake vigorously from side to side for 20 min.
7. Let sit, undisturbed, for at least 1 h.

Observation:

(A) Indole test: Red color indicates positive MR (*E. coli*); yellow or orange color indicates negative MR (*Klebsiella*).

(B) MR-VP tests: Pink color indicates positive VP (*Klebsiella*); no pink color indicates negative VP (*E. coli*).

Applications:

1. For differentiation of the various members of *Enterobacteriaceae*.
2. In water purification system.

Chapter 38

To Perform Biochemical Identification of Microorganism by Nitrogen Source Utilization or Urease Test

Principle: Urea agar medium contains urea and phenol red. Urease is an enzyme that catalyzes the conversion of urea to CO_2 and NH_3. Ammonia combines with water to produce ammonium hydroxide, a strong base which increases pH of the medium. Increase in the pH causes phenol red to turn deep pink. This is indicative of a positive reaction for urease.

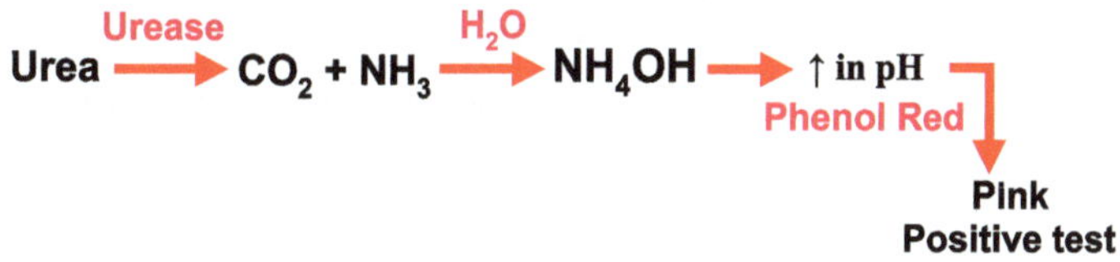

Procedure:

- Streak a urea agar tube with the organism (*Escherichia coli, Enterobacter aerogenes, Proteus Vulgaris, Pseudomonas aeruginosa*).
- Incubate at 37 °C for 24 h.

Observation:

- If color of medium turns from yellow to pink, it indicates positive test.
- *Proteus* gives positive reaction after 4 h, while *Klebsiella* and *Enterobacter* gave positive results after 24 h.

Aakanchha Jain et al. *Basic Techniques in Biochemistry, Microbiology and Molecular Biology: Principles and Techniques*, Springer Protocols Handbooks, https://doi.org/10.1007/978-1-4939-9861-6_38,

Viva Voce:

Q1. Which organism gives red color in indole test?

Q2. What is the principle of indole test?

Q3. What is the principle of MR VP test?

Q4. Which test is used for differentiation of *Klebsiella* from *E. Coli?*

Q5. Give the composition of Barritt's A reagent.

Q6. Give the composition of Barritt's B reagent.

Q7. Give the composition of Kovac's reagent.

Q8. Name few microbes belonging to *Enterobacteriaceae* family.

Q9. Enzymatic degradation of tryptophan forms.......................

Q10. Give two application of IMViC test?

Q11. Give the principle of urease test.

Q12. Which color is obtained in positive urease test?

Q13. What is the incubation time to obtain positive test for *Klebsiella?*

Q14. What is the incubation time to obtain positive test for *Enterobacter?*

Chapter 39

Immobilization of Enzymes and Microorganisms

1 Enzyme Immobilization in Polyacrylamide Gel

Principle: Immobilization in polyacrylamide gel is a technique based on the polymerization of acrylamide with N,N′-methylene-bis-acrylamide (Bis) as the cross-linking agent. The degree of cross-linking is controlled by adjusting the ratio of acrylamide to Bis used. Being nonionic in nature, polyacrylamide is the most widely used matrix for entrapping enzymes.

Equipment: Beakers, Pipettes, Balance, Graduated cylinder, Syringe, and Needle

Materials:

Solution A (pH 7.0)

- 0.1 mM EDTA
- 0.1 M Tris-HCI
- 1.1% N,N′-Methylene-bis-acrylamide and 20% acrylamide

Washing solution (pH 7.0)

- 0.5 M NaCl
- 0.1 mM EDTA
- 0.1 M Tris-HCI

Dimethylaminopropionitrile (polymerization initiator)

Potassium persulfate solution, 1% (polymerization catalyst)

Nitrogen gas cylinder

7.5% fungal amylase (or any enzyme having optimum activity at pH 7)

Aakanchha Jain et al. *Basic Techniques in Biochemistry, Microbiology and Molecular Biology: Principles and Techniques*, Springer Protocols Handbooks, https://doi.org/10.1007/978-1-4939-9861-6_39,

Procedure:

1. Prepare solution "A" by dissolving 1.1 g of N,N′-Methylene-bis-acrylamide and 20 g of acrylamide to a 100 ml of solution (pH 7.0) having 0.1 mM EDTA and 0.1 M Tris-HCl as buffer.
2. Add enzyme to 10 ml of the solution "A" prepared in step 1 and mix.
3. For 20 min, aerate the solution with oxygen for 20 min.
4. Add 0.1 ml of dimethylaminopropionitrile, mixing continuously. Add 1.0 ml of freshly prepared 10% potassium persulfate solution to start the polymerization.
5. Immediately pour the solution in molds to avoid gel formation in beaker. This will take nearly 20–30 min
6. Alternatively, if smaller pieces are desired, small molds can be used.
7. Wash the free enzyme 2–3 times from gel surface using wash bottle filled with washing solution.

2 Enzyme Immobilization in Alginate Gel

Principle: When a proper ration of calcium as salt and alginate in its sodium salt, i.e., in ionic form is present, replacement reaction causes formation of calcium alginate which thickens and acquires the spherical shape called beads. Alternatively the gel can be set in molds of desired sizes.

$$2\ \text{Na Alginate} + CaCl_2 = \text{Ca (Alginate)}2 + 2\ \text{Na}+ \quad (1)$$

The ratio smaller or equal to 25:1 of sodium-calcium gives a stable gel and bead without tailing. Calcium chloride should be used preferably at 3 mM concentration.

Equipments: Beakers, Graduated cylinder, Balance, Pipettes, Syringe

Materials:

Sodium alginate (5% w/v)

Sodium chloride (0.9% w/v)

Calcium chloride (0.5 M)

Enzyme

Procedure:

1. Prepare 3% w/v solution of sodium alginate in distilled water by constant heating.

2. Add 0.015 g of enzyme in warm not hot solution with 10 ml of sodium alginate solution.
3. Fill the solution in needle-less syringe. Take 75 ml of calcium chloride solution in a 100 ml measuring cylinder.
4. Add the sodium alginate solution in the calcium chloride solution dropwise. Beads will be formed. Allow curing of beads for 1 h.
5. Wash the beads with sodium chloride solution (0.9% w/v) and store in the same for further use.

Precaution: After sodium alginate is completely dissolved, leave the solution undisturbed for 30 min to eliminate the air bubbles that can later be entrapped and cause the beads to float.

3 Enzyme Immobilization in Gelatin Gel

Principle: The process is simply based on hydrogel property of gelatin. It effectively involves hardening of gel to provide structural strength.

Equipments: Beakers, Graduated cylinder, Pipettes, Constant temperature bath Freezer

Materials:

Gelatin Solution (10% w/v)

Hardening solution

Formaldehyde (20% v/v)

Ethanol (50% v/v)

Water (30% v/v)

Enzyme Extract

Procedure:

1. Prepare 100 ml of gelatin solution in water (10% w/v) by heating the solution at 35–40 °C. The enzyme chosen should be stable at this temperature and show optimum activity.
2. Add formaldehyde as gelling agent. 0.015 g of the enzyme per 10 ml of the solution is added; the gel is left to set by pouring in molds (cylindrical molds/beads/flat membranes/cubes) of desired size and shape.
3. Freeze at −28 °C for 2 h increases the gelling process.
4. Wash the gel 2–3 times with deionized water using wash bottles.

4 Cell Immobilization in Agarose

Principle: Agarose a linear, neutral polysaccharide dissolves in water at elevated temperature. On cooling the gel is formed. Agarose preparations with a gelling temperature between 28 and 40 °C are employed for the immobilization of microbial and plant cells.

Procedure:

1. 2 g fresh weight plant cells or 2 ml microbial suspension are dispersed in 10 ml, 1–4% w/w agarose maintained at 40 °C.
2. 50 ml of vegetable oil is also maintained at 40 °C.
3. Add the suspension in step 1 to the oil. Droplets will be seen.
4. Cool the mixture at 15 °C on an ice bath under continuous stirring until the agarose beads solidify.
5. Buffer is added to settle down the beads under gravity.
6. The oil and aqueous phase is removed by pipette.
7. The beads are washed with buffer until the smell of organic phase disappears.
8. Sonication can be used to reduce the bead size (1–10 nm).

5 Immobilization of Microbial Cells in Gelatin

Materials: Deionized water, ethanol, silica gel, gelatin, formaldehyde, microbial cells

Procedure:

1. Microbial cells are suspended in deionized water and dispersed in a 10% w/v gelatin aqueous solution at 35–40 °C to give a final cell to gelatin ratio of 1:10 on a dry weight basis.
2. To 9.5 ml of this dispersion 0.5 ml of a hardening solution consisting of 20% ww formaldehyde in 50% v/v ethanol is added. The mixture is poured into a cylinder (0.8 cm diameter) and kept in a deep freezer (about −25 °C).
3. After at least 2–3 hours gelatin cylinder obtained is thoroughly rinsed with tap water, kept over in a large volume of deionized water at refrigerator temperature (4–5 °C), and then cut into thin disks (0.2–04 cm). The gel disks can be stored in refrigerator for months without loss of enzyme activity (e.g., yeast cell invertase). Else it can be stored wet in deionized water containing a proper preservative or dry in a desiccator over silica gel.

Viva Voce:

Q1. ....% of N,N′-Methylene-bis-acrylamide and ....% Acrylamide are used for enzyme immobilization.

Q2. State whether true or false.

(i) 5% gelatin solution is used for immobilization...........

(ii) 1–2% agarose forms good gel for entrapment of enzyme...........

(iii) Calcium alginate beads are example of covalent immobilization.......

(iv) In cell immobilization sucrose, glucose are used as osmoticum with calcium salt..........

(v) Sonication can be used to reduce bead size...........

Q3. Define immobilization.

Q4. Which polymer is used for cell immobilization.

Q5. ....cross-linking agent is used in preparation of calcium alginate beads.

Part IV

Biochemistry

Chapter 40

Qualitative Systematic Analysis of Carbohydrates (Glucose, Fructose, Lactose, Maltose, Sucrose and Starch)

Carbohydrates are polyhydroxyaldehyde or ketone derivatives of polyhydric alcohols with the general chemical formula $Cn(H_2O)n$, and these are largely distributed in animals and plants. Carbohydrates provide energy through oxidation and supplies carbon for the synthesis of cell components; they serve as a form of stored chemical energy. Carbohydrates, along with lipids, proteins, nucleic acids, and other compounds are known as biomolecules because they are closely associated with living organisms (Table 1).

1 Molisch's Test (α-Naphthol Reaction)

Principle: Monosaccharide is treated with concentrated sulfuric acid (H_2SO_4) or concentrated hydrochloric acid (HCl), -OH group of sugar are removed in the form of water, and furfural is formed from pentose sugar, and hydroxymethylfurfural is formed from hexose sugar. These products react with sulfonated α-naphthol and give a purple (violet-red)-colored ring at junction of two liquids.

Reagent Preparation: 5% solution of α-naphthol in alcohol.

Procedures: Add 2 drops of Molisch's reagent to the 5 mL of sugar solution in the test tube. Mix gently, and then add 2 mL H_2SO_4 (conc.) by the side of the slanted test tube. The formation of violet-red layer appears at the junction. Acid goes down below the sugar solution as it is heavier (Fig. 1).

Aakanchha Jain et al. *Basic Techniques in Biochemistry, Microbiology and Molecular Biology: Principles and Techniques*, Springer Protocols Handbooks, https://doi.org/10.1007/978-1-4939-9861-6_40,

Table 1
Classification of carbohydrates

Type	Sugar/sugar units	Example
Monosaccharides	Simple sugar	Glucose, fructose, mannose, galactose, etc.
Disaccharides	2 units of simple sugar	Sucrose, maltose, lactose
Oligosaccharides	3–10 units of simple sugar	Maltotriose, raffinose
Polysaccharide	More than 10 units of simple sugar	Starch, insulin, cellulose, glycogen

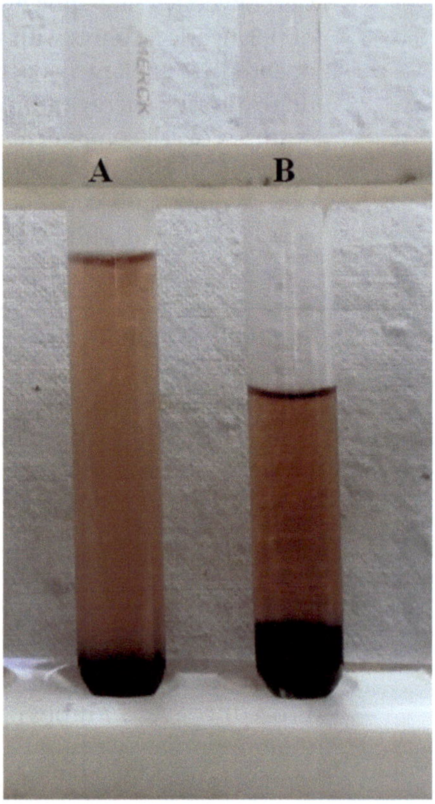

Fig. 1 Molisch's test

Observations:

1. Purple (violet-red) color appears at the junction of two liquids.
2. A brown color due to charring must be ignored, and the test should be repeated with a more dilute sugar solution.
3. Green color appears due to impurities or addition of excess Molisch's reagent.

Uses and Applications: Generally used for the detection of carbohydrates.

2 Benedict's Test

Principle: Carbohydrates with a free aldehyde or keto group have the ability to reduce copper sulfate into cuprous oxide due to formation of enediols in the alkaline medium when Benedict's reagent is used.

Reagent: Dissolve 17.3 g of copper sulfate in 100 mL of water. Dissolve 173 g of sodium citrate and 100 g of sodium carbonate in water by continuous heating. To this, add slowly copper sulfate solution, and allow cooling the solution by transferring in the liter flask.

Procedure: Add 8 drops of the sugar solution to the 5 mL of Benedict's reagent. Vigorously boil the solution for 2 min. (or on water bath for 3 min). Keep the test tube to cool in rack (do not cool quickly in cold water) (Fig. 2).

Observation: Color may vary depending on the concentration of sugar (orange-red green or yellow color).
Test does not result positive if solution contains less than 0.1–0.15 g% of sugar (Table 2).

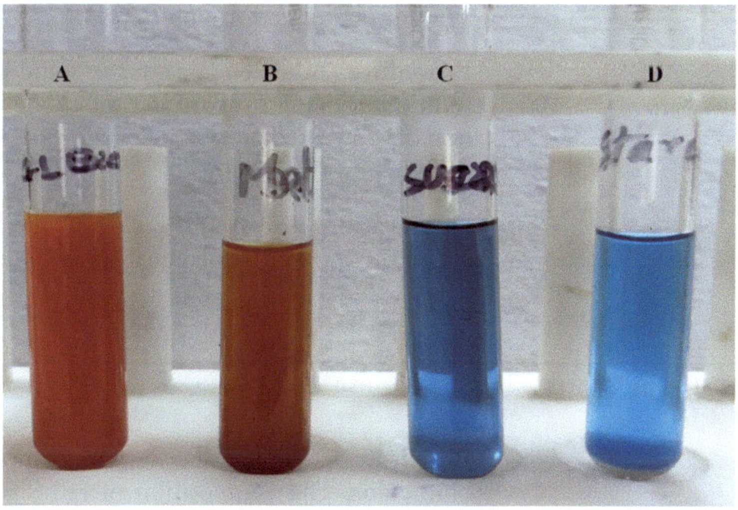

Fig. 2 Benedict's test. (**a**) Glucose, (**b**) fructose, (**c**) starch, (**d**) sucrose

Table 2
Different colored precipitates observed from Benedict's test according to sugar concentration

S. no.	Observation after treatment with Benedict's reagent	Conclusion	Glucose (mg/dl)
1	Blue	Sugar absent	–
2	Green to yellow	Traces of sugar present	250–500
3	Green to dark yellow	Sugar present	500–1000
4	Yellow to orange	Sugar present	1000–1500
5	Orange to red	Sugar present	Greater than 1500

3 Fehling's Test

Principle: Carbohydrates when heated with Fehling's solution gives reddish brown precipitate due to reduction of copper sulfate to cuprous oxide indicating presence of reducing sugar.

Reagent: Fehling's reagents comprise of two solutions, Fehling's solution A and solution B. Fehling's solution A is prepared by dissolving copper sulfate (6.93 g) in water (100 mL), and Fehling's solution B is prepared by dissolving potassium hydroxide (20 g) and sodium potassium tartrate (34.6 g) in water (100 mL). Rochelle salts (sodium potassium tartrate) present in the reagent acts as the chelating agent. Both solutions are mixed in equal amount before performing the test.

Procedure: Add 2 mL of sugar solution to 5 mL of Fehling's solution in the test tube. Heat the tube on water bath. The formation of brown-red precipitate of cuprous oxide appears at the end of the reaction (Fig. 3).

Observation: Brown-red precipitate of cuprous oxide appears indicating presence of reducing sugar.

4 Iodine Test

Principle: Starch forms adsorption complex with iodine and gives blue color. The composition of red-brown or blue color is not defined.

Reagent: Iodine is dissolved in an aqueous solution of potassium iodide, and 0.5 mL of iodine is diluted up to 5 mL using distilled water.

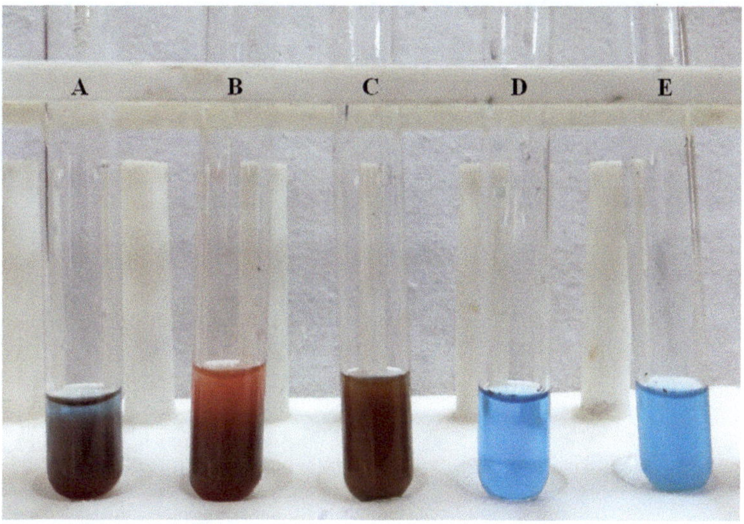

Fig. 3 Fehling's test. (**a**) Lactose, (**b**) fructose, (**c**) glucose, (**d**) sucrose, (**e**) starch

Procedure: Add 2 drops of iodine solution (0.05 N) to 2–3 mL of polysaccharide solution (starch, dextrin, and glycogen).

Observation: Different colors appear for each of the polysaccharide solution. In case of starch, dextrin, and glycogen, red-brown or deep blue color appears, while blue color disappears on heating in case of starch and color comeback on subsequent cooling. Color does not reappear in case of dextrin and glycogen (Fig. 4).

Questions:

Explain the reaction between iodine solution and polysaccharides by giving the structures of related compounds.

Each polysaccharide tested gives different color results with the iodine test. Explain the reason briefly.

5 Nylander's Test (Bismuth Reduction Test)

Principle: Free ketone or aldehyde group of carbohydrate reduces to form black precipitates. Even very small amount of glucose (0.08%) can be detected by this test.

Reagent: Dissolve 4 g of Rochelle salt and 2 g of bismuth subnitrate in 10% solution (100 mL) of potassium hydroxide.

Procedure: Take 5 ml sample solution in a test tube, add 5–8 drops of Nylander's reagent, boil for 3 min on water bath, and cool.

Observation: Black precipitates develop after a few minutes (bismuth subnitrate reduced to black bismuth).

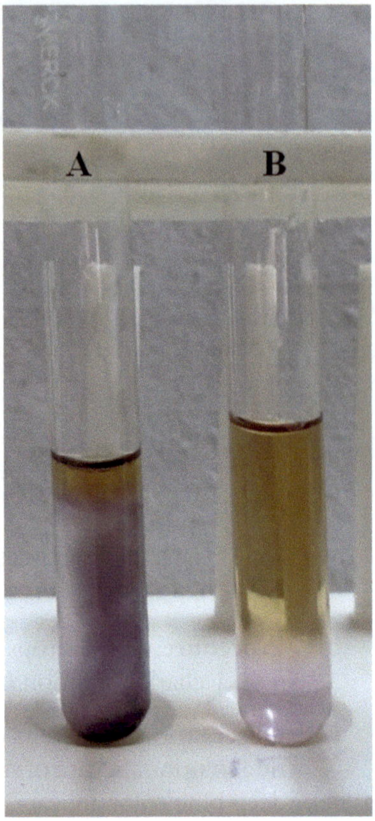

Fig. 4 Iodine test. (**a**) Glucose, (**b**) Starch

6 Cole's Test

Principle: Free ketone or aldehyde group of carbohydrate reduces copper sulfate to cuprous oxide and forms yellow precipitate.

Procedure: Add a large quantity of anhydrous sodium carbonate, 3 drops of 5% copper sulfate, and 3 drops of glycerol to 5 mL 0.1% test solution and boil.

Observation: Formation of yellow precipitate on boiling confirms the presence of carbohydrate.

7 Reduction of Methylene Blue Test

Principle: Free ketone or aldehyde group of carbohydrate reduces methylene blue to leuco-methylene blue and gives blue color.

Procedure: Take 3 ml distilled water and 0.5 ml 40% NaOH solution in a test tube, add a drop of 1% methylene blue, boil, and cool. Take care that the color of solution remains blue. Add 1 ml of 0.2% sugar solution, boil again, and cool.

Observation: Blue color of the solution disappears (methylene blue reduced to leuco-methylene blue- colorless).

8 Barfoed's Test: Reducing Monosaccharide

Principle: Barfoed's test reaction is based on the reduction of cupric acetate by reducing monosaccharides and reducing disaccharides. The free aldehyde and ketone groups of monosaccharide reduce copper sulfate to cuprous oxide and give red precipitates.

Reagent: To 450 mL of boiling water, add 24 g of copper acetate. Do not filter if precipitate forms, and add 8.5% lactic acid (25 mL) to the above hot solution. Cool the solution, and make the volume up to 500 mL and filter.

Procedure: Take 5 ml Barfoed's reagent in a test tube, add 8 drops of sugar solution, and stir and heat to boiling for 1 min. Allow to stand for at least 15 min.

Observation: Orange or red-colored precipitates of cuprous oxide appear within 5 min indicating presence of monosaccharide, while it takes 7–12 min for disaccharides (Fig. 5).

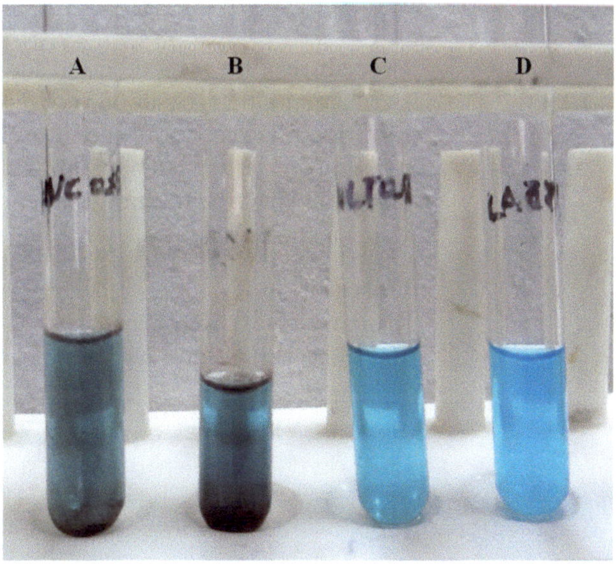

Fig. 5 Barfoed's test. (**a**) Glucose, (**b**) Fructose, (**c**) Maltose, (**d**) Lactose

9 Bial's Test for Pentose

Principle: Hydrochloric acid when reacted with pentose sugar (such as ribose sugar) forms furfural derivative, and then reaction with orcinol in the presence of ferric ion forms a blue-green color compound.

Reagent: Bial's reagent is prepared by dissolving 150 mg of orcinol to 50 mL of concentrated hydrochloric acid, and add 3 drops of ferric chloride solution (10%).

Procedure: Two ml of a sample solution is placed in a test tube. Two ml of Bial's reagent is added. The solution is then heated gently in a Bunsen burner or hot water bath. If the color is not evident, more water can be added to the tube.

Observation: Formation of blue color indicates presence of pentose. Other colors indicate a negative result for pentoses. Hexoses usually react to form green, red, or brown products.

10 Seliwanoff's Test

Keto sugars get dried in the concentrated acids to yield furfurals or their subsidiaries which react with resorcinol in Seliwanoff's reagent to yield a cherry-red complex.

Reagent: The reagent is prepared by dissolving 50 mg of resorcinol in 35 mL of concentrated HCL and diluted up to 100 mL with water.

Procedure: Take a test tube and add 5 ml of Seliwanoff's reagent in it. Make sure that the amount of reagent you are adding in the test tube doesn't exceed 5 ml and measure it before pouring it. Now for the sample to be tested, measure 1 ml of it and pour it down in the test tube. Heat the solution using boiling water for 5 min and then wait for the results.

Observation: A cherry-red color shows the presence of ketoses (fructose sugar). Ketopentoses yield blue-green color, while aldoses and disaccharides give no shading change (Fig. 6).

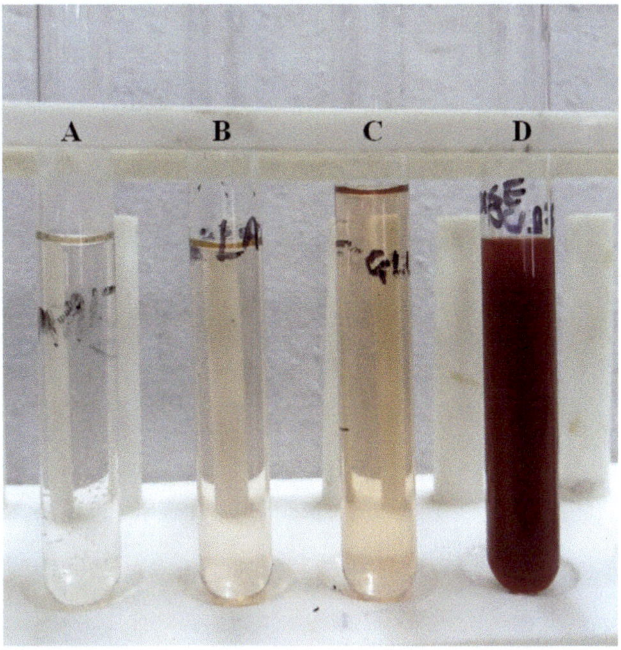

Fig. 6 Seliwanoff's test. (**a**) Maltose, (**b**) Lactose, (**c**) Glucose, (**d**) Fructose

11 Hydrolysis of Sucrose

Principle: Sucrose upon hydrolysis with HCl gives fructose and glucose.

Procedure: Take a test tube and add 5 ml of sucrose solution. To this solution add one drop of thymol solution and two drops of HCl. Take half of the solution in another test tube. Boil the solution of one tube and cool it under tap water. To neutralize the pH of both, add dropwise Na_2CO_3 solution (2%). The pink acidic solution upon neutralization gives blue color. The boiled sample is hydrolyzed to give fructose and glucose and now can be detected by Benedict's and Seliwanoff's test. The second unboiled solution does not reduce Benedict's solution.

12 Methylamine Test for Lactose

Reagent: Methylamine hydrochloride (0.2%), sodium hydroxide (10%).

Procedure: In a test tube, take 5 ml of sample sugar solution, and add 1 mL of methylamine hydrochloride (0.2%) followed by sodium hydroxide (10%). Mix the solution properly by covering the tube with bulb, and heat the solution at 56 °C for 30 min and then keep it at room temperature. If the lactose is present in much amount, red color is observed before the end of the heating.

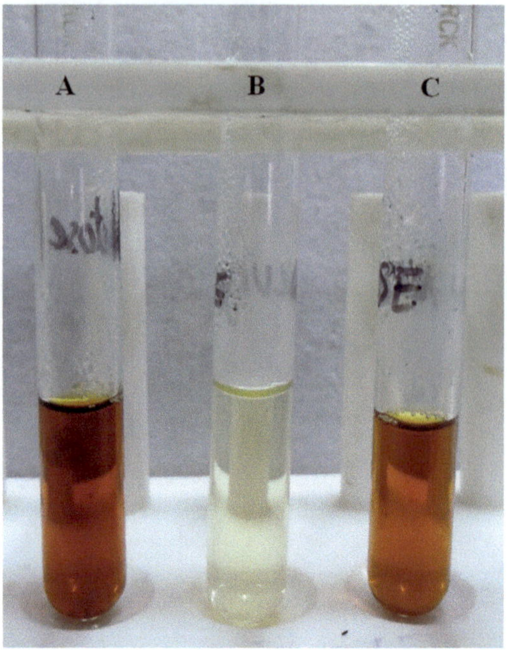

Fig. 7 Methylamine test. (**a**) Lactose, (**b**) Glucose, (**c**) Maltose

Observation: Disaccharides, lactose, and maltose give red color, while yellow color is observed if galactose, glucose, fructose, xylose, and sucrose are present in much amount (Fig. 7).

13 Osazone Test

Principle: Monosaccharide reacts with phenylhydrazine which is a crystalline compound. The sugars which will reduce as an effect of this reaction will result in the formation of osazones. Simple sugars like glucose, fructose, and mannose produce the same osazone because of molecular structure similarity.

Procedure: To 5 mL of sugar solution, add 10 drops of glacial acetic acid and then 0.5 g of phenylhydrazine hydrochloride, and double the amount of acetate crystals and then mix well. Keep the tube in a boiling water bath for 20 min, and then allow it to cool slowly. Carefully examine the deposits under a microscope (Fig. 8a–e).

Discussion: Osazone crystal formation depends on time schedule (Tables 3 and 4).

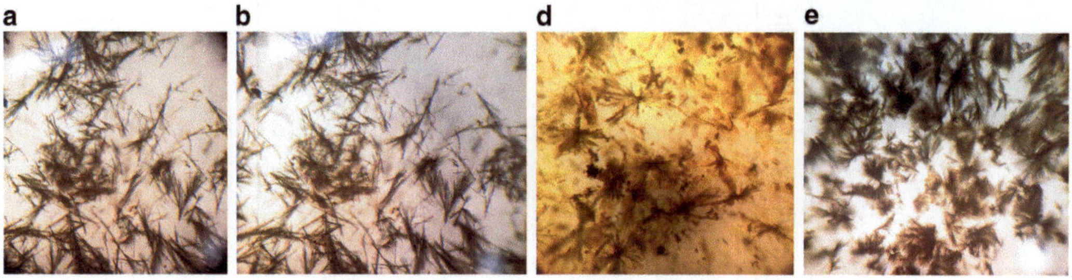

Fig. 8 (**a**) Glucosazone, (**b**) Fructosazone, (**c**) Galactosazone, (**d**) Maltosazone, (**e**) Lactosazone

Table 3
Shapes of crystal seen from different osazones

Osazone	Shape of crystal seen
Glucosazone	Broomstick-like
Fructosazone	Broomstick-like
Galactosazone	Rhombic-like
Lactosazone	Powderpuff-like
Maltosazone	Sunflower-like

Viva Voce:

Q1. What is the composition of Molisch's reagent?

Q2. Which product is formed in the Fehling's test of reducing sugars?

Q3. What is Fehling's solution A?

Q4. What is Fehling's solution B?

Q5. Why do glucose, mannose, and fructose give similar osazone crystals?

Q6. Give the difference between Benedict's and Barfoed's test.

Q7. Examples of reducing disaccharides.

Q8. Why is it that the Benedict's test and Barfoed's test are not suitable for testing glucose in urine?

Q9. The boiling step is common for each test for the reducing sugars. Why boiling is necessary for the reduction to take place?

Q10. Explain the difference between glucose solution and alkali-treated glucose solution when a test for reducing sugars is applied.

Table 4
Observation of carbohydrate test results

Test	Carbohydrate	Glucose	Fructose	Galactose	Pentose	Maltose	lactose	Sucrose	Starch	glycogen	dextrin
Molisch's test											
Iodine test											
Benedict's test											
Fehling test											
Barfoed's test											
Bial's test											
Tauber's test											
Seliwanoff's test											
Hydrolysis test											
Methylamine test											
Osazone test											

Put + or – signs for positive or negative response

Chapter 41

Protein Analysis in Food

Proteins are an abundant component in all cells, and almost all except storage proteins are important for biological functions and cell structure. Food proteins are very complex. Many have been purified and characterized. Proteins can be classified by their composition, structure, biological function, or solubility properties. For example, simple proteins contain only amino acids upon hydrolysis, but conjugated proteins also contain nonamino acid components. The solubility as well as functional properties of proteins could be altered by denaturants. The analysis of proteins is complicated by the fact that some food components possess similar physicochemical properties. Numerous methods have been developed to measure protein content. The basic principles of these methods include the determinations of nitrogen, peptide bonds, aromatic amino acids, dye-binding capacity, ultraviolet absorptivity, and light scattering properties.

Protein analysis is required when it is necessary to know the total protein content, protein content of a specific protein in a mixture, and during isolation and purification of a protein, nonprotein nitrogen, amino acid composition, and nutritive value of a protein.

Analysis of food protein is not essentially a direct procedure. This is due to foods being heterogenic materials, covering a range of different nutrients, such as lipids, carbohydrates, and micronutrients. The composition, food structure, or matrix and interactions between the different nutrients may reduce the ease of access of the proteins, leading to underestimation of the protein content.

Food Sources: *Protein content in food varies widely. Foods of animal source and pulses are tremendous sources of proteins.*

Aakanchha Jain et al. *Basic Techniques in Biochemistry, Microbiology and Molecular Biology: Principles and Techniques*, Springer Protocols Handbooks, https://doi.org/10.1007/978-1-4939-9861-6_41,

1 Kjeldahl Method

Principle: In the Kjeldahl method, proteins and other organic food components in a sample in the presence of catalyst are digested with sulfuric acid. The total organic nitrogen is changed to ammonium sulfate. The digest is neutralized with alkali and distilled into a boric acid solution. The borate anions formed are titrated with standardized acid, which is converted to nitrogen in the sample. The result of the analysis represents the crude protein content of the food since nitrogen also comes from nonprotein components.

Procedure: Sample Preparation

20-mesh-size solid powder is used for the study, so that they are homogeneous in size.

Digestion: Place sample (accurately weighed) in a Kjeldahl flask. Add acid and catalyst; digest until clear to get complete breakdown of all organic matter. Nonvolatile ammonium sulfate is formed from the reaction of nitrogen and sulfuric acid:

$$\text{Protein} = \frac{\text{Sulfuric acid}}{\text{Heat, Catalyst}} \times (NH_4)_2SO_4 \tag{1}$$

During digestion, protein nitrogen is liberated to form ammonium ions; sulfuric acid oxidizes organic matter and combines with ammonium formed; and carbon and hydrogen elements are converted to carbon dioxide and water.

Neutralization and Distillation: The digest is diluted with water. Alkali-containing sodium thiosulfate is added to neutralize the sulfuric acid. The ammonia formed is distilled into a boric acid solution containing the indicators methylene blue and methyl red (AOAC Method 991.20):

$$(NH_4)_2SO_4 + 2NaOH \rightarrow 2NH_3 + Na_2SO_4 + 2H_2O \tag{2}$$

$$NH_3 + H_3BO_3(\text{Boric acid}) \rightarrow NH_4 + H_2BO_3^- \tag{3}$$

Titration: Borate anion (proportional to the amount of nitrogen) is titrated with standardized HCl.

Calculations:

$$\text{Moles of HCl} = \text{moles } NH_3 = \text{moles} Y \text{ in the sample}$$

A reagent blank should be run to subtract reagent nitrogen from the sample nitrogen:

$$\%Y \text{ N HCL} \times \frac{\text{Corrected acid volume}}{\text{g of sample}} \times \frac{14 \text{ g N}}{\text{mole}} \times 100 \tag{4}$$

where:

N HCl = normality of HCl in moles/1000 mL

Corrected acid vol. = (mL std. acid for sample) – (mL std. acid for blank)

14 = atomic weight of nitrogen

A factor is used to convert percent N to percent crude protein. Most proteins contain 16% N, so the conversion factor is 6.25 (100/16 = 6.25):

$$\frac{\%Y}{0.16} = \%\text{protein} \qquad (5)$$

Applications: The Kjeldahl method is an AOAC official method for crude protein content and has been the basis for evaluation of many other protein methods. The Kjeldahl method is still used for some applications, but nowadays its use is limited in many countries due to the availability and advantages of automated nitrogen combustion (Dumas) systems.

2 Dumas (Nitrogen Combustion)

Method

Principle: The combustion method was introduced in 1831 by Jean-Baptiste Dumas. It has been modified and automated to improve accuracy since that time. Samples are combusted at high temperatures (700–1000 °C) with a flow of pure oxygen. All carbon in the sample is converted to carbon dioxide during the flash combustion. Nitrogen-containing components produced include N2 and nitrogen oxides. Nitrogen oxides are reduced to nitrogen in a copper reduction column at a high temperature (600 °C). The total nitrogen (including inorganic fraction, i.e., nitrate and nitrite) released is carried by pure helium and quantitated by gas chromatography using a thermal conductivity detector (TCD). Ultrahigh purity acetanilide and EDTA (ethylenediaminetetraacetate) may be used as the standards for the calibration of the nitrogen analyzer. The nitrogen determined is converted to protein content in the sample using a protein conversion factor.

Procedure: Samples (approximately 100–500 mg) are weighed into a tin capsule and introduced to a combustion reactor in automated equipment. The nitrogen released is measured by a built-in gas chromatograph.

Applications: The combustion method is a faster and safer alternative to the Kjeldahl method and is suitable for all types of foods. As an AOAC method, the Dumas method is widely used for official purposes, but its speed also allows for quality control applications. The industry uses different units/systems, depending on sample size and protein content. Freeze-drying can be used to concentrate diluted liquid samples, e.g., waste steam samples.

3 Infrared Spectroscopy

Principle: Infrared spectroscopy measures the absorption of radiation (near- or mid-infrared regions) by molecules in food or other substances. Different functional groups in a food absorb different frequencies of radiation. For proteins and peptides, various mid-infrared bands (6.47 μm) and near-infrared (NIR) bands (e.g., 3300–3500 nm, 2080–2220 nm, 1560–1670 nm) characteristic of the peptide bond can be used to estimate the protein content of a food. By irradiating a sample with a wavelength of infrared light specific for the constituent to be measured, it is possible to predict the concentration of that constituent by measuring the energy that is reflected or transmitted by the sample (which is inversely proportional to the energy absorbed).

Applications: Mid-infrared spectroscopy is used in infrared milk analyzers to determine milk protein content, while near-infrared spectroscopy is applicable to a wide range of food products (e.g., grains, cereal, meat, and dairy products), especially as a rapid method to test nonstandard milk.

4 Colorimetric Methods

When protein reacts with specific reagents under certain conditions, colorful compounds are generated, and the absorbance is measured by spectrophotometer. The protein content is expressed on the basis of standard protein such as bovine serum albumin (BSA), and thus this method is not an absolute method. Due to the differences in the composition of proteins, these methods have limited use. However, because of high sensitivity, these methods have the advantage of requiring a small sample size.

5 Dye-Binding Methods

5.1 Anionic Dye-Binding Method

Principle: The protein-containing sample is mixed with a known excess amount of anionic dye in a buffered solution. Proteins bind the dye to form an insoluble complex. The unbound soluble dye is measured after equilibration of the reaction and the removal of insoluble complex by centrifugation or filtration. Excess insoluble complex dye is separated using anionic sulfonic acid dye, including acid orange12, orange G, and Amido Black 10B, binds cationic groups of the **basic amino acid residues** (imidazole of histidine, guanidine of arginine, and ε-amino group of lysine) and the **free amino-terminal group** of the protein. The amount of the unbound dye is inversely related to the protein content of the sample.

Procedure:

1. The sample is finely ground (60 mesh or smaller sizes) and added to an excess dye solution with known concentration.
2. The content is vigorously shaken to equilibrate the dye-binding reactions and filtered or centrifuged to remove insoluble substances.
3. Absorbance of the unbound dye solution in the filtrate or supernatant is measured and dye concentration estimated from a dye standard curve.
4. A straight calibration curve can be obtained by plotting the unbound dye concentration against total nitrogen (as determined by Kjeldahl method) of a given food covering a wide range of protein content.
5. Protein content of the unknown sample of the same food type can be estimated from the calibration curve or from a regression equation calculated by the least squares method.

Applications: Anionic dye-binding has been used to estimate proteins in milk, wheat flour, soy products, and meats. The anionic dye-binding method may be used to estimate the changes in available lysine content of cereal products during processing since the dye does not bind altered, unavailable lysine. Since lysine is the limiting amino acid in cereal products, the available lysine content represents protein nutritive value of the cereal products.

5.2 Bradford Dye-Binding Method

Principle: When Coomassie Brilliant Blue G-250 binds to protein, the dye changes color from reddish to bluish, and the absorption maximum of the dye is shifted from 465 to595 nm. The change in the absorbance at 595 nm is proportional to the protein concentration of the sample. Like other dye-binding methods, the Bradford method relies on the amphoteric nature of proteins. When the protein-containing solution is acidified to a pH less than the isoelectric point of the protein(s) of interest, the dye added binds electrostatically. Binding efficiency is enhanced by hydrophobic interaction of the dye molecule with the polypeptide backbone adjoining positively charged residues in the protein. In the case of the Bradford method, the dye bound to protein has a change in absorbance spectrum relative to the unbound dye.

Procedure:

1. Coomassie Brilliant Blue G-250 is dissolved in 95% ethanol and acidified with 85% phosphoric acid.
2. Samples containing proteins (1–100 μg/mL) and standard BSA solutions are mixed with the Bradford reagent.
3. Absorbance at 595 nm is read against a reagent blank.
4. Protein concentration in the sample is estimated from the BSA standard curve.

Applications: The Bradford method has been used successfully to determine protein content in worts and beer products and in potato tubers. This procedure has been improved to measure microgram quantities of proteins. Due to its rapidity, sensitivity, and fewer interferences than the Lowry method, the Bradford method has been used widely for the analysis of low concentrations of proteins and enzymes in their purification and characterizations.

6 Copper Ion-Based Methods

Since the biuret method was first developed to measure protein content based on the reaction with copper ions, several improved methods have been developed. Following the description of the basic biuret method below are the modified Lowry and the bicinchoninic acid (BCA) methods, which are both based in part on the biuret method.

6.1 Biuret Method

Principle: A violet-purplish color is produced when cupric ions are complexed with peptide bonds (substances containing at least two peptide bonds, i.e., oligopeptides, large peptides, and all proteins) under alkaline conditions. The absorbance of the color produced is read at 540 nm. The color intensity (absorbance) is proportional to the protein content of the sample.

Procedure:

1. A 5 mL biuret reagent is mixed with a 1 mL portion of protein solution (1–10 mg protein/mL).The reagent includes copper sulfate, NaOH, and potassium sodium tartrate, which is used to stabilize the cupric ion in the alkaline solution.
2. After the reaction mix is allowed to stand at room temperature for 15 or 30 min, the absorbance is read at 540 nm against a reagent blank. Filtration or centrifugation before reading absorbance is required if the reaction mixture is not clear. A standard curve of concentration versus absorbance is constructed using bovine serum albumin (BSA).

Applications: The biuret method has been used to determine proteins in cereal, meat, and soybean proteins and as a qualitative test for animal feed. The biuret method also can be used to measure the protein content of isolated proteins.

6.2 Lowry Method

Principle: The Lowry method combines the biuret reaction with the reduction of the Folin-Ciocalteu phenol reagent (phosphomolybdic-phosphotungstic acid) bytyrosine and tryptophan residues in the proteins. The bluish color developed is read at 750 nm(high sensitivity for low protein concentration) or 500 nm (low sensitivity for high protein concentration).The original procedure has been modified by Miller and Hartree to improve the linearity of the color response to protein concentration and replace the use of two unstable reagents with one stable reagent.

Procedure: The following procedure is based on the modified procedure of Hartree:

1. Proteins to be analyzed are diluted to an appropriate range (20–100 μg).
2. K Na tartrate-Na_2CO_3 solution is added after cooling and then incubated at room temperature for 10 min.
3. $CuSO_4$-K Na tartrate-NaOH solution is added after cooling and then incubated at room temperature for 10 min.
4. Freshly prepared Folin's reagent is added, and then the reaction mixture is mixed and incubated at 50 °C for 10 min.
5. Absorbance is read at 650 nm.
6. A standard curve of BSA is carefully constructed for estimating protein concentration of the unknown.

Applications: Because of its simplicity and sensitivity, the Lowry method has been widely used in protein biochemistry. However, it has not been widely used to determine proteins in food systems without first extracting the proteins from the food mixture.

6.3 Bicinchoninic Acid Method

Principle: Proteins and peptides (as short as dipeptides) reduce cupric ions to cuprous ions under alkaline conditions which is similar in principle to that of the biuret reaction. The cuprous ion then reacts with the apple-greenish bicinchoninic acid reagent (BCA) to form a purplish complex (one cuprous ion is chelated by two BCA molecules). The color measured at 562 nm is nearly linearly proportional to protein concentration over a wide range of concentration from micrograms up to 2 mg/mL. Peptide bonds and four amino acids (cysteine, cystine, tryptophan, and tyrosine) contribute to the color formation with BCA.

Procedure:

1. Mix (one step) the protein solution with the BCA reagent, which contains BCA sodium salt,sodium carbonate, NaOH, and copper sulfate, pH 11.25.
2. Incubate at 37 °C for 30 min or room temperature for 2 h or 60 °C for 30 min. The selection of the temperature depends upon the sensitivity desired. A higher temperature gives a greater color response.
3. Read the solution at 562 nm against a reagent blank.
4. Construct a standard curve using BSA.

Applications: The BCA method is widely used in protein isolation and purification due to its advantage (over the modified Lowry method and any Coomassie dye-based assay) of being compatible with samples containing up to 5% detergents. While most dye-binding methods are faster, the BCA method is less affected by protein compositional differences, so there is better protein-to-protein uniformity.

7 Ultraviolet Absorption at 280 nm

Principle: Proteins show strong absorption in the ultraviolet (UV) region at UV 280 nm, primarily due to tryptophan and tyrosine residues in the proteins. Because the content of tryptophan and tyrosine in proteins from each food source is fairly constant, the absorbance at 280 nm could be used to estimate the concentration of proteins, using Beer's law. Since each protein has a unique aromatic amino acid composition, the extinction coefficient (E280) or molar absorptivity (Em) must be determined for individual proteins for protein content estimation.

Procedure:

1. Proteins are solubilized in buffer or alkali.
2. Absorbance of protein solution is read at 280 nm against a reagent blank.
3. Protein concentration is calculated according to the following equation:

$$A = abc$$

where A = absorbance; a = absorptivity; b = cell or cuvette path length; and c = concentration.

Applications: The UV 280 nm method has been used to determine the protein contents of milk and meat products. It has not been used widely in food systems. This technique is better applied in a purified protein system or to proteins that have been extracted in alkali or denaturing agents such as 8 M urea. Although peptide bonds in proteins absorb more strongly at 190–220 nm than at 280 nm, the low UV region is more difficult to measure.

8 Peptide Measurement at 190–220 nm

Peptides without or with low level of tyrosine or tryptophan residues can be quantified at 190–220 nm at which peptide bonds have maximum absorption. The extinction coefficients in the far UV range can be calculated with the consideration of contribution of tyrosine and tryptophan to the absorption. Protein can also be measured in this UV range.

Viva Voce:

Q1. Why is the protein analysis necessary?

Q2. What is the role of sulfuric acid in Kjeldahl method?

Q3. Which method is suitable for crude protein content and why?

Q4. What is the alternative of Kjeldahl method?

Q5. Give the example of anionic dye.

Q6. What is the principle of Bradford dye-binding method?

Q7. Which method is used for the estimation of isolated proteins?

Q8. What is the principle of nitrogen combustion method?

Q9. How is IR spectroscopy used for determining protein content?

Q10. Give some examples of dye-based protein assay.

Chapter 42

Identification Tests for Proteins (Casein and Albumin)

Casein: Casein, like proteins, is made up of many hundreds of individual amino acids. Each may have a positive or a negative charge, depending on the pH of the [milk] system. At some pH value, all the positive charges and all the negative charges on the [casein] protein will be in balance, so that the net charge on the protein will be zero. That pH value is known as the isoelectric point (IEP) of the protein and is generally the pH at which the protein is least soluble. For casein, the IEP is approximately 4.6, and it is the pH value at which acid casein is precipitated. In milk, which has a pH of about 6.6, the casein micelles have a net negative charge and are quite stable. During the addition of acid to milk, the negative charges on the outer surface of the micelle are neutralized (the phosphate groups are protonated), and the neutral protein precipitates.

The same principle applies when milk is fermented to curd. The lactic acid bacillus produces lactic acid as the major metabolic end product of carbohydrate [lactose in milk] fermentation. The lactic acid production lowers the pH of milk to the IEP of casein. At this pH, casein precipitates.

Casein is precipitated from milk at pH 4.6 using acetic acid and sodium acetate solutions in Kjeldahl flask. The acidified solution which contains the non-casein nitrogen components is separated from casein precipitate by filtration. Nitrogen content of the casein precipitate is determined by Kjeldahl method and multiplied by 6.38 to obtain casein in protein.

Reagents:

- Sodium acetate solution—1 M/liter using AR sodium acetate or sodium acetate trihydrate—Transfer 4.10 ± 0.1 g sodium acetate or 6.80 ± 0.1 g sodium acetate trihydrate into 50 ml volumetric flask, and dilute to volume with water. Prepare fresh weekly.
- Acetic acid solution—10% using AR grade glacial acetic acid.

Aakanchha Jain et al. *Basic Techniques in Biochemistry, Microbiology and Molecular Biology: Principles and Techniques*, Springer Protocols Handbooks, https://doi.org/10.1007/978-1-4939-9861-6_42,

- Buffer solution—Dilute 1 ± 0.1 ml sodium acetate and 1 ± 0.1 ml 10 5 acetic acid to 100 ml with water. Prepare fresh weekly.

Procedure (Kjeldahl Method): Weigh 1–2 g sample into Kjeldahl flask and add 50 ml water. Add 0.75 ml 10% acetic acid to flask and swirl gently. Leave mixture for 10 min. Add 0.75 ml of sodium acetate and swirl gently. Pour mixture from Kjeldahl flask on to pleated filter paper (Whatman No 1, 15 cm) and collect filtrate. Let it drain completely before next pour. Add 30 ml of buffer solution to Kjeldahl flask and swirl to mix. Pour mixture onto filter paper after first filtration is complete and combine filtrates. Add another 30 ml of buffer solution and add filtrate to the previous two filtrates. Filtrate should be clear and free of particulate matter. If particulates appear recycle filtrate through the same filter paper or repeat test. Remove filter paper. Ensure no precipitate is lost on filter paper. Drop filter paper into Kjeldahl flask, add pot sulfate and copper sulfate and sulfuric acid, and digest. Determine nitrogen content and multiply by 6.38 to obtain casein protein.

Albumin: Albumins are compact and roughly spherical in shape and have axial ratios not more than 3 (that is the ratio of their shortest to longest dimensions). Hence albumins come under globular proteins. They are soluble with definite molecular weight. Albumins of interest are serum albumin of blood, lactalbumin of milk, and ovalbumin of egg. It is also present in pulses. They are soluble in solute-free water and coagulable on heating. They are not precipitated by half saturation.

Biuret Test:

Principle: Biuret test is a general test for compounds having a peptide bond. Biuret is a compound formed by heating urea to 180 °C. When biuret is treated with dilute copper sulfate in alkaline condition, a purple-colored compound is formed. This is the basis of biuret test widely used for identification of proteins and amino acids.

The principle of biuret test is conveniently used to detect the presence of proteins in biological fluids. Alkaline $CuSO_4$ reacts with compounds containing two or more peptide bonds to give a violet-colored product which is due to formation of coordination complex of cupric ions with unshared electron pairs of peptide nitrogen and O_2 of water.

Reagents:

- Biuret reagent, i.e., mixture of hydrated copper sulfate, potassium hydroxide solution, and potassium sodium tartrate.

- 1% alanine and 5% egg white (albumin) (positive control).
- Deionized water (negative control).

Procedure: Take three clean and dry test tubes. Add 1–2 ml of the test solution, egg albumin, and deionized water in the respective test tubes. Add 1–2 ml of Biuret reagent to all the test tubes. Shake well and allow the mixtures to stand for 5 min. Observe for any color change.

Note: care must be taken that not more than two drops of dilute copper sulfate (1%) be added; otherwise blue color will develop instead of violet.

Interpretation: A yellow precipitate is formed.

Ninhydrin Test:

This test is due to a reaction between an amino group of free amino acid and ninhydrin. Ninhydrin is a powerful oxidizing agent, in its presence, amino acid undergo oxidative deamination liberating ammonia, CO_2, a corresponding aldehyde and reduced form of ninhydrin. The NH_3 formed from an amino group reacts with another molecule of ninhydrin and its reduced product (hydrindantin) to give a blue substance diketohydrin (Ruhemanns).

Procedure: Take 1 ml test solution in dry test tube and 1 ml distilled water in another tube as a control. Pour few drops of 2% ninhydrin in both the test tubes. Keep the test tubes in water bath for 5 min. Look for the development of blue or violet color.

Interpretation: Intense blue color is formed.

Millon's Test:

Principle: The mercurous and mercuric nitrate reacts with the hydroxybenzene radicals (phenols) forming a red-colored compound. In other words, Millon's reagent reacts with phenolic group of tyrosine to form mercuric fumarate which gives pinkish- or red-colored compound.

Reagents:

- Test solution: 1% arginine, 1% tyrosine, phenol solution
- Millon's reagent (acidified mercuric sulfate)
- 1% sodium nitrite

Procedure: Take 1 ml test solution in dry test tube. Similarly, take 1 ml distilled water in another test tube as control. Add 1 ml of Millon's reagent and mix well. Boil gently for 1 min. Cool under tap water. Now add five drops of 1% sodium nitrite. Heat the solution slightly. Look for the development of brick-red precipitate.

Interpretation: White precipitate changes to brick red on boiling.

Viva Voce:

Q1. What is casein?

Q2. What is the nature of casein?

Q3. What is isoelectric pH of casein?

Q4. What is the range of albumin concentration of plasma?

Q5. What is the composition of biuret reagent?

Q6. How to determine nitrogen content in casein?

Q7. What is the principle of ninhydrin test?

Q8. What is the reason of violet color instead of blue color in biuret test?

Q9. What is the effect on albumin concentration in liver disease?

Q10. Give an example of carrier protein which helps in transportation of biomolecule.

Chapter 43

Quantitative Analysis of Reducing Sugars by 3, 5-Dinitrosalicylic Acid (DNSA Method)

Principle: The method is based on the detection of presence of free carbonyl C=O group of reducing sugars. Initially oxidation of the ketonic and aldehyde functional group of fructose and glucose, respectively, by 3, 5-Dinitrosalicylic acid (yellow color) to by 3-amino-5-nitrosalicylic acid (orange-red)* in alkaline medium.

Materials and Reagents:

1. 250 μg/mL of 100 mL standard reducing sugar solution (fructose, glucose, maltose)
2. 1% w/v, 60 mL of stock reducing sugar standard in saturated benzoic acid
3. 2 mol/lit sodium hydroxide
4. Sodium potassium tartrate solution (300 g in 500 mL distilled water)/Rochelle salt solution
5. 3, 5-Dinitrosalicylic acid solution (10 g in 200 mL 2 mol/lit sodium hydroxide)
6. Freshly prepared dinitrosalicylic acid reagent [DNS] (mix solutions 4 and 5, and make up the volume to 1 l with water just before use).
7. Boiling water bath
8. 0.5% w/v sodium sulfate (used to absorb dissolved oxygen which may interfere in oxidation of glucose)
9. 2.0% w/v crystalline phenol (used as color intensifier)

Procedures:

1. Add 1 mL of DNSA reagent to 3 mL of reducing sugar sample in screw-capped tubes.
2. Blank solution is prepared by adding 1 mL of DNS reagent to 3 mL of distilled water.

Aakanchha Jain et al. *Basic Techniques in Biochemistry, Microbiology and Molecular Biology: Principles and Techniques*, Springer Protocols Handbooks, https://doi.org/10.1007/978-1-4939-9861-6_43,

Table 1
Standard curve of reducing sugar

S. no.	Concentration (μg/mL)	Absorbance (nm) at λmax 540 nm
1.	10	
2.	20	
3.	30	
4.	40	
5.	50	
6.	60	
7.	70	

Absorbance of unknown sample of reducing sugar = ...

3. Place the tubes in boiling water bath for 5 min and cool at room temperature. A red-brown color develops.
4. Add 1 mL of sodium sulfate solution to obtain a stable color.
5. Read the absorbance of tube at 540 nm against blank.
6. The standard curve may be prepared and used to determine the concentration of reducing sugar in unknown samples (Table 1).

Applications: For quantitative analysis of sugars in newly prepared ready-to-eat food items, sweetened pharmaceutical preparation; in detection of reducing sugars in the process of fermentation.

Viva Voce:

Q1. What is the full form of DNSA?

Q2. The absorbance for quantitation of DNA is taken at . . .. lambda max.

Q3. What is the use of sodium sulfate in DNSA method?

Q4. What are reducing sugars?

Q5. What are the other qualitative tests for detection of reducing sugars?

Q6. Give the application of quantitative analysis of reducing sugars by 3, 5-dinitrosalicylic acid.

Q7. Why is the method not suitable for combination of sugars?

Q8. Which chemical is used as color intensifier?

Q9. Give the principle of DNSA method.

Q10. What percentage of sugar is used for preparation of standard stock solution in DNSA method?

Note: Different sugars give unlike colors. Thus, method is not utilized for combination of reducing sugars.

Chapter 44

Quantitative Analysis of Proteins by Various Methods Including Biuret

Principle: Biuret test is a general test for compounds having a peptide bond. Biuret is a compound formed by heating urea to 180 °C. When biuret is treated with dilute copper sulfate in alkaline condition, a purple-colored compound is formed. This is the basis of biuret test widely used for identification of proteins and amino acids.

The principle of biuret test is conveniently used to detect the presence of proteins in biological fluids. Alkaline $CuSO_4$ reacts with compounds containing two or more peptide bonds to give a violet-colored product which is due to formation of coordination complex of cupric ions with unshared electron pairs of peptide nitrogen and O_2 of water.

Reagents:

- Biuret reagent, i.e., mixture of hydrated copper sulfate, potassium hydroxide solution, and potassium sodium tartrate.
- 1% alanine, 5% egg white (albumin) (positive control).
- Deionized water (negative control).

Procedure: Take three clean and dry test tubes. Add 1–2 ml of the test solution, egg albumin, and deionized water in the respective test tubes. Add 1–2 ml of biuret reagent to all the test tubes. Shake well and allow the mixtures to stand for 5 min. Observe for any color change.

Note: Care must be taken that not more than two drops of dilute copper sulfate (1%) be added; otherwise blue color will develop instead of violet (Table 1).

Aakanchha Jain et al. *Basic Techniques in Biochemistry, Microbiology and Molecular Biology: Principles and Techniques*, Springer Protocols Handbooks, https://doi.org/10.1007/978-1-4939-9861-6_44,

Table 1
Observation table of Biuret test

S. no.	Observation	Interpretation
1.	No color change (solution remains blue)	Proteins are not present
2.	The solution turns from blue to violet (deep purple)	Proteins are present
3.	The solution turns from blue to pink	Peptides are present (Peptides or peptones are short chains of amino acid residues which contain less number of peptide bonds)

1 Other Methods

1.1 Bradford Method

Principle: The protein in solution can be measured quantitatively by different methods. The methods described by Bradford uses a different concept—the protein's capacity to bind to a dye, quantitatively. The assay is based on the ability of proteins to bind to Coomassie brilliant blue and form a complex whose extinction coefficient is much greater than that of free dye.

Reagents:

- Dissolve 100 mg of Coomassie Brilliant Blue G-250 in 50 ml of 95% ethanol.
- Add 100 ml of 85% phosphoric acid and make up to 600 ml with distilled water.
- Filter the solution and add 100 ml of glycerol and then make up to 1000 ml.
- The solution can be used after 24 h.
- BSA.

Procedure: Prepare various concentrations of standard protein solutions from the stock solution (say 0.2, 0.4, 0.6, 0.8, and 1.0 ml) into series of test tubes, and make up the volume to 1 ml. Pipette out 0.2 ml of the sample in two other test tubes, and make up the volume to 1 ml. A tube with 1 ml of water serves as blank. Add 5.0 ml of Coomassie brilliant blue to each tube, and mix by vortex or inversion. Wait for 10–30 min and read each of the standards and each of the samples at 595 nm. Plot the absorbance of the standards versus their concentration. Plot the graph of optical density versus concentration. From the graph find the amount of protein in unknown sample (Table 2).

Calculations:

Result: The amount of protein present in the given sample was found to be ___________

Table 2
Observation for Barfoed's method

S.No.	Vol. of BSA (ml)	Concentration of BSA (mg/ml)	Volume of distilled water (ml)	Volume of Bradford reagent (ml)	Incubation for 10 min	OD at 595 nm

1.2 Lowry Method

Principle: The principle behind the Lowry method of determining protein concentrations lies in the reactivity of the peptide nitrogen with the copper[II] ions under alkaline conditions and the subsequent reduction of the Folin-Ciocalteu phosphomolybdic phosphotungstic acid to heteropolymolybdenum blue by the copper-catalyzed oxidation of aromatic acids. The Lowry method is sensitive to pH changes, and therefore the pH of assay solution should be maintained at 10–10.5.

Reagents:

2% Na_2CO_3 in 0.1 N NaOH.

1% NaK tartrate in H_2O.

0.5% $CuSO_4.5\ H_2O$ in H_2O.

Reagent I: 48 ml of A, 1 ml of B, and 1 ml C.

Reagent II: 1 part Folin phenol [2 N]:1 part water.

Bovine serum albumin standard: 1 mg/ ml.

Procedure: 0.2 ml of BSA is used as working standard in five test tubes and volume made up to 1 ml using distilled water. The test tube with 1 ml distilled water serves as blank. Add 4.5 ml of reagent I and incubate for 10 min. After incubation add 0.5 ml of reagent II and incubate for 30 min. Measure the absorbance at 660 nm and plot the standard graph. Estimate the amount of protein present in the given sample from the standard graph (Table 3).

Calculations:

Result: The amount of protein present in the given sample was found to be ___________

Table 3
Observation for Lowry method

S.No.	Vol.of BSA (ml)	Concentration of BSA (mg/ml)	Volume of distilled water (ml)	Volume of reagent I (ml)	Incubation for 10 min	Volume of reagent II (ml)	Incubation for 10 min	OD at 660 nm

1.3 Protein Precipitation

Principle: Proteins are precipitated by salts of the heavy metals (e.g., $HgCl_2$, $AgNO_3$, etc.). Proteins are usually soluble in water solutions, because they have hydrophobic amino acids on their surfaces that attract water molecules and interact with them. This solubility is a function of the ionic strength and pH of the solution. Proteins have isoelectric points at which the charges of their amino acid side groups balance each other. If the ionic strength of a solution is either very high or very low, proteins will tend to precipitate at their isoelectric point. The solubility is also a function of ionic strength, and as you increase the ionic strength by adding salt, proteins will precipitate. Ammonium sulfate is the most common salt used for this purpose because it is unusually soluble in cold buffers (we have to keep proteins cold!) and is economically viable.

1.3.1 Ammonium Sulfate Precipitation

Reagents:

Ammonium sulfate.

BSA.

Alkaline copper sulfate solution.

Folin's reagent.

Procedure: Dissolve 5 mg of BSA in 10 ml of distilled water in 50 ml beaker. Add the corresponding saturation amount (50–100%) of ammonium sulfate to the beaker with stirring. After adding ammonium sulfate completely to the beaker, leave it undisturbed for 30 min. Transfer the solution to the 15 ml centrifuge tube, and centrifuge at 10000 rpm for 5 min. Discard the supernatant, and dissolve the pellet in distilled water and centrifuge again. Discard the supernatant, and dissolve the pellet in 1 ml of distilled water. Take 200 μl of the sample, and estimate the amount of protein present by Lowry method. Calculate the percentage of recovery by using the formula (Table 4):

Table 4
Observation for protein precipitation method

Percentage of ammonium sulfate	Protein concentration	% recovery
50		
60		
70		
80		
90		
100		

$$\text{Recovery percentage} = \frac{\text{Final protein concentration}}{\text{Initial protein concentration}} \times 100 \qquad (1)$$

Result: Maximum percentage recovery of BSA is_________________at _____________ % of ammonium sulfate saturation.

Viva Voce:

Q1. Which test is suitable for the compounds having peptide bonds?

Q2. What is the composition of biuret reagent?

Q3. What is the principle of Bradford method?

Q4. What is the pH required for assay using Lowry method?

Q5. How will you determine the amount of protein in unknown sample?

Q6. Give examples of heavy metals used for precipitation of protein.

Q7. Why is biuret test give a violet color?

Q8. What is the principle of protein precipitation?

Q9. Name the reagent used for the precipitation of protein.

Q10. Name the reagents used for and their role in Lowery method.

Chapter 45

Qualitative Analysis of Urine for Abnormal Constituents

1 Physical Properties

Volume: In ancient period analysis of body excretion was done for diagnosis of diseased condition. Generally fresh urine of normal person is transparent with aromatic odor-containing urochrome. Normal adult individuals excrete urine in a range of 1.2–1.8 l per day. Increased output of urine called as polyuria arises in diseased conditions like contracted kidney, diabetes, etc., while reduction in output (oliguria) is observed in certain shock and nephritis. Anuria, complete suppression of urine, occurs in renal failure.

Properties of normal urine:

Average volume: 1200 ml

Specific activity: 1.003–1.030

Average pH: 6.0

Total solids: 30–70 g/liter

Biochemical Composition of Urine (Released in 24 h):

Sodium: 3–4 gm

Creatinine: 60–150 mg

Potassium: 1.5–2.0 gm

Iodine: 50–250 μg

Arsenic: less than 50 μg

Chloride: 9–16 gm

Inorganic phosphorus: 1–1.5 gm

Magnesium: 0.05–0.2 gm

Aakanchha Jain et al. *Basic Techniques in Biochemistry, Microbiology and Molecular Biology: Principles and Techniques*, Springer Protocols Handbooks, https://doi.org/10.1007/978-1-4939-9861-6_45,

Ketone bodies: 3–15 mg

Ammonia: 0.3–1.0 gm

Lead: less than 50 μg

Urea: 25–30 gm

Creatinine: 1.0–1.8 gm

Calcium: 0.1–0.3 gm

Uric acid: 0.3–1.0 gm

Hippuric acid: 0.1–1.0 gm

Sulfur: 0.7–3.5 gm

Purine bases: 7–10 mg

Oxalic acid: 15–20 mg

Indican: 4–2 mg

Allantoin: 20–30 mg

Coprophyrins: 60–280 μg/ml

Phenols: 0.2–0.5 gm

Vitamins, hormones, and enzymes: traces

Glucose, pentoses, and lactose: 0.01%

Physical Characteristics of Urine: Parameters: volume, color, appearance, odor, sediment formation (due to amorphous urates, phosphates, leukocytes, epithelial cells.) reaction, pH, and specific gravity.

Determination of Specific Gravity: It is ratio of weight of a volume of urine to weight of the same volume of distilled water at specific temperature. Determination of specific gravity helps to measure diluting and concentrating power of kidney.

Materials: Wide mouth glass or plastic bottle (250 ml capacity) with screw cap, urinometer, thermometer, pipette, filter paper, and pH strip.

Procedure:

1. Collect first voided midstream morning urine sample in glass/ plastic bottle.
2. Place in urinometer and float the hydrometer on urine sample and note down the reading by hydrometer.
3. Measure temperature of urine.

Observations: If temperature of urine differs from that mentioned on hydrometer, correct readings as:

(a) Add 0.001 for each 3 °C above the stated temperature.

(b) Subtract 0.001 for each 3 °C below the stated temperature.

Note:

- Hydrometer should not touch the sides of vessel.
- Qualitative test: morning urine sample is preferred.
- Quantitative test: 24 h urine sample is preferred.
- Sample should be processed within 1 h for better results.

2 Biochemical Characterization of Urine to Detect Abnormal Constituents

Carbohydrates: In diseased condition like diabetes mellitus, increase in level of glucose appears in the urine. Depending upon sugar appearing in urine, the conditions are termed as pentosuria (pentoses), fructosuria (fructose), and galactosuria (galactose). Benedict's test is one of the most commonly used methods to detect presence of reducing sugar in urine sample.

3 Benedict's Test

Materials Required: Clean glass tube, glass pipette, water bath, Benedict's reagent, and test tube stand.

Procedure:

1. Take 1–2 ml of urine in clean glass test tube.
2. To it add 5 ml of Benedict's reagent (commercially available), and boil the mixture for 5–10 min.
3. The color of mixture changes to yellow/red depending on the amount of sugar.
4. The change in color confirms presence of reducing sugars in it (Fig. 1) of Chap. 40 and Fig. 1 below.

Fig. 1 Benedict's test in urine sample

4 Tauber's Test for Detection of Aldopentoses in Urine Samples

Materials: Benzidine solution, 2% acetic acid, water bath, clean glass test tubes, pipette, and test tube stand.

Procedure:

- Take 0.1 ml of urine sample in test tube, and add 0.5 ml of benzidine solution (2% in glacial acetic acid); boil the tubes for 5 min in water bath.
- Allow the tubes to cool at room temperature; add 1 ml of distilled water.
- Control tube: take 0.1 ml of distilled water in test tube and add 0.5 ml of benzidine solution (2% in glacial acetic acid); boil the tubes for 5 min in water bath.
- Formation of pink to red color indicates the presence of an aldopentose.

5 Bial's Test for Detection of Pentoses in Urine Samples

Materials: Bial's reagent (0.2% concentrated hydrochloric acid containing 0.25 mL of 10% ferric chloride), water bath, clean glass test tubes, pipette, and test tube stand.

Procedure:

- Pipette 2 ml of urine in test tube, and add 5 ml of Bial's reagent.
- Heat the tube in boiling water bath and cool.
- Appearance of green color indicates presence of pentoses.

6 Seliwanoff's Test for Detection of D-Fructose in Urine Samples

HCl reacts with fructose to form furfuraldehyde which forms red colored compound on reaction with resorcinol.

Materials: Seliwanoff's reagent (mix 50 mg resorcinol in 33 ml of conc. HCL, make up the volume to 100 ml), water bath, clean glass test tubes, pipette, and test tube stand.

Procedure:

1. Take 0.5 ml of urine samples and add 5 ml of Seliwanoff's reagent in clean glass test tube.
2. Boil the mixture for 5 min in water bath.
3. Observe the color of mixture: Red color indicates presence of fructose, and no change in color indicates negative (fructose absent) result.

7 Mucic Acid Test for Analysis of Lactose or Galactose

Lactose and galactose form an insoluble acid on oxidation with nitric acid.

Procedure:

1. In a clean glass beaker, take 40 ml urine sample and to it dropwise add conc. HNO_3.
2. Heat the mixture in water bath to reduce the volume to one-fourth of total mixture.
3. After cooling add water and leave overnight.
4. Observation: Formation of white precipitate confirms presence of lactose and galactose.

8 Detection of Acetone Bodies Using Rothera's Nitroprusside Test

Materials: Urine sample, water bath, clean glass test tubes, pipette, test tube stand, ammonium sulfate, sodium nitroprusside solution (5%), and ammonium hydroxide.

Procedure:

1. Take 5 ml of urine sample in clean glass tube, and saturate it by ammonium sulfate.
2. To the mixture add 2–3 drops of freshly prepared 5% sodium nitroprusside solution (freshly prepared), mix well, and add 1–2 ml of ammonium hydroxide solution.
3. Observation: the appearance of permanganate color denotes the presence of acetone.

9 Estimation of Acetoacetic Acid by Gerhardt's Test

Materials: Urine sample, water bath, clean glass test tubes, pipette, test tube stand, filter paper, and 5% ferric chloride solution.

Procedure:

1. Pipette 5 mL of urine sample in a test tube, and dropwise add 5% of ferric chloride solution.
2. Add until no further precipitation appears.
3. Filter the precipitate with ordinary filter paper.
4. To the filtrate further add ferric chloride solution.
5. Observation: development of red color confirms presence of acetoacetic acid.

Note: Antipyrine, salicylates, and phenacetin show similar observations.

10 Analysis of Proteins in Urine Sample by Sulfosalicylic Acid Test

Materials: Urine sample, water bath, clean glass test tubes, pipette, test tube stand, filter paper, sulfosalicylic acid (20%), and sodium sulfate (20%).

Procedure:

1. Take 2 ml of urine sample in clean test tube; dropwise (3–5drops) add sulfosalicylic acid (20%) and sodium sulfate (20%).
2. Observation: two layers are formed and at the interface white layer of precipitated protein is visible.

Fig. 2 Bence Jones test in urine sample

11 Analysis of Proteins in Urine Sample by Bence Jones Protein

Cancerous tumor forms Bence Jones protein which is building block of the antibodies. Thus detecting this protein in urine helps doctors to diagnose multiple myeloma.

Materials: Urine sample, water bath, clean glass test tubes, pipette, test tube stand, filter paper, and 3% sulfosalicylic acid.

Procedure:

1. Take 10 ml of fresh urine sample in test tube, and acidify it with 3% sulfosalicylic acid.
2. When the pH of mixture reaches 5.0, boil tubes in water bath for 5 min. Filter the hot mixture with ordinary filter paper.
3. Observation: Appearance of precipitate or turbidity on cooling indicates positive test results. Further heating shows disappearance of precipitate confirms presence of Bence Jones protein (Fig. 2).

12 Determination of Bile Pigments by Harrison Spot Test

Barium chloride reacts with sulfur radicals present in urine, thereby forming barium sulfate precipitate. Bile pigments present in urine adhere to this precipitate. On adding Fouchet's reagent, bilirubin forms green-colored compound biliverdin.

Materials: Urine sample, water bath, clean glass test tubes, centrifuge tubes, Pasteur pipette, test tube stand, filter paper, 10 g/dl barium chloride, Fouchet's reagent (10% solution of ferric chloride, 10 ml; trichloroacetic acid,25 g; distilled water, 100 ml), and Ehrlich's reagent (2% solution of p-dimethyl amino benzaldehyde in 50% HCl).

Procedure:

1. Take 10 ml of urine in centrifuge tube, and to it add 10 ml of $BaCl_2$(10%); mix well.
2. Centrifuge the mixture at 1500 rpm for 8–10 min.
3. Filter the supernatant, and add 0.5 ml of Ehrlich's reagent to supernatant.
4. Add 3–5 drops of Fouchet's reagent to sediment (pellet).

Observations: Appearance of green color in sediment indicates presence of bile salts.

Development of cherry red color in supernatant indicates presence of urobilinogen.

13 Schlesinger's Test for Urobilinogen

Materials: Urine sample, water bath, clean glass test tubes, centrifuge tubes, Pasteur pipette, test tube stand, filter paper, calcium chloride solution (10%), Lugol's solution (4 gm iodine and 6.0 gm potassium iodide in 100 mL of distilled water), zinc chloride, or zinc acetate.

Procedure:

Part A:

Removal of bile pigment

1. *To 10 ml of urine sample, add 2 ml $CaCl_2$ solution (10%, w/v) and mix well.*
2. *Remove precipitated calcium salts of the bile pigments.*
3. *Use filtrate for Schlesinger's test.*

Part B:

Schlesinger's test

a. *Dispense 10 ml urine sample in test tube, and add 3–5 drops of Lugol's solution.*
b. *Add 10 ml of saturated alcoholic solution of zinc chloride.*
c. *Observation: Formation of a green fluorescence shows the presence of urobilin (an oxidized product of urobilinogen) or urobilinogen.*

14 Detection of Bile Salts in Urine by Pettenkofer's Test

Materials: Urine sample, water bath, clean glass test tubes, centrifuge tubes, Pasteur pipette, test tube stand, filter paper, cane sugar, and conc. H_2SO_4.

Procedure:

1. Add crystals of cane sugar (10–15) to 5 mL of urine sample in a test tube; mix well.
2. To mixture add 3 ml of conc. H_2SO_4 from wall of test tube.
3. Observation: a red to violet ring which is formed at junction of two layers indicates the presence of bile salts.

15 Detection of Occult Blood in Urine

Materials: Urine sample, water bath, clean glass test tubes, centrifuge tubes, Pasteur pipette, test tube stand, filter paper, benzidine, glacial acetic acid, H_2O_2 (3% v/v), and orthotoulidine.

Procedure:

A. **Benzidine Test**

1. Take 2 ml of urine in test tube.
2. Add 3 ml of benzidine (saturated solution in glacial acetic acid) to it.
3. Add 1 ml of H_2O_2 (3% v/v).
4. Observation: Appearance of a blue or green color indicates a positive result (Fig. 3).

Note: Positive and negative control should be prepared as reference.

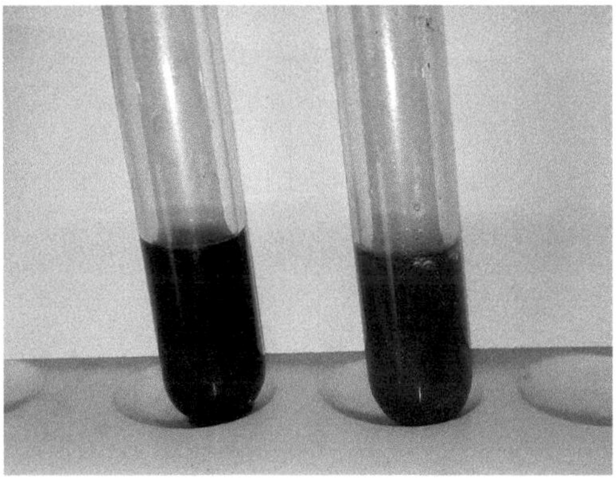

Fig. 3 Benzidine test in urine sample

B. **Orthotoulidine Test**

1. Take 1 ml of urine sample in clean glass tube; add equal amount of orthotoulidine solution in glacial acetic acid (4% w/v).
2. To mixture add 1 mL of 3% hydrogen peroxide solution.
3. Observation: The formation of blue color indicates presence of occult blood.

Viva Voce:

Q1. Give the examples of solid component of urine in normal health.

Q2. What is glucosuria? Enlist some common conditions of glycosuria.

Q3. Why the bile salts reduces the surface tension of urine?

Q4. What is the principle of Rothera's test?

Q5. What are conditions in which ketone bodies are formed?

Q6. What is the principle of Hay's test?

Q7. Give the example of abnormal constituents of urine.

Q8. Name the test specifically used for the detection of urea in urine.

Q9. Give the example of nonnitrogenous constituents of urine.

Q10. What are the chief nitrogenous constituents of urine?

Chapter 46

Determination of Blood Creatinine

Creatinine tests are used to estimate the waste product "creatinine," in blood and urine. It is produced due to metabolism of muscle tissue. Creatinine test is used to monitor kidney disorders (renal), e.g., acute and chronic renal failure, and also to measure the effect on kidney due to other diseases (e.g., heart, liver, etc.). The body produces creatinine everyday constantly. It is formed when creatine is converted into energy. Creatinine is a by-product of that process. The creatinine is excreted from the body if kidney functioning is normal and accumulates in blood stream when kidney functioning impaired. Synthesis of creatine occurs in the kidney and liver which then moves in circulation and is taken up by the skeletal muscles. Only about 2% of creatine is converted to creatinine daily. Normal serum creatinine level in male ranges 0.7–1.5 mg/dl and in females 0.4–1.2 mg/dl. Only serum is preferred and deproteinized before test because RBCs are rich in non-creatinine chromogens (e.g., glucose, urea, uric acid, proteins, etc.).

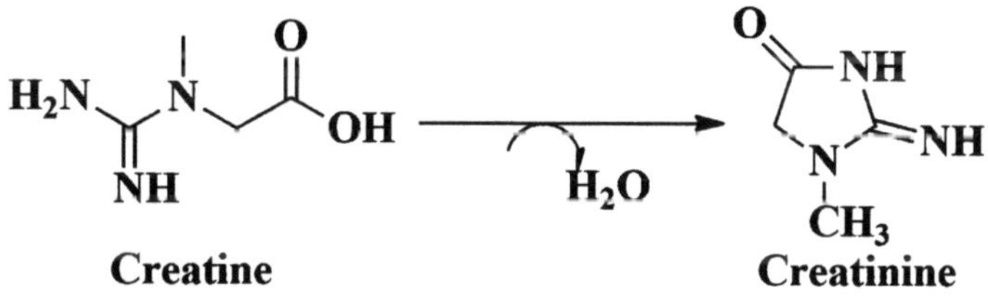

Aakanchha Jain et al. *Basic Techniques in Biochemistry, Microbiology and Molecular Biology: Principles and Techniques*,
Springer Protocols Handbooks, https://doi.org/10.1007/978-1-4939-9861-6_46,

1 Serum Creatinine Test (Jaffe's Test)

Principle: Creatinine reacts with picrate in alkaline medium and develops a red-orange color complex creatinine picrate. The intensity of color produced from the sample can be measured by colorimetric analysis at wavelength of 520 nm. The intensity of color is directly proportional to the amount of creatinine present.

$$\text{Creatinine} + \text{Pciric Acid in Alkaline medium} \\ = \text{Creatinine Picrate (Red color Jaffe's complex)}$$

Reagents: Picric acid (0.04 M), sodium hydroxide (5%), sulfuric acid (2/3 M), creatinine standard (100 mg %), sodium tungstate solution (10%).

Procedure:

1. Take 2 mL serum and 2 mL distilled water in centrifuge tube.
2. Add 2 mL sodium tungstate to precipitate the protein.
3. Add 0.6 N sulfuric (2 mL) with constant shaking.
4. Keep for 10 min and filter the solution.
5. The obtained protein-free filtrate is used for test.
6. Take three rest tubes, and mark them as blank (B), standard (S), and test (T), and arrange the test tube in the order given in Table 1.

Preparation of Standard Curve:

1. Prepared the different concentration of creatinine in the test tube and prepare one as blank. Make the final volume up to 2 mL.
2. Add 1 mL of NaOH solution to each tube and stir.
3. Add 1 mL picric acid, mix, and keep aside for 10 min at room temperature.

Table 1
Setup of experiment

Addition sequence	B (mL)	S (mL)	T (mL)
Supernatant	–	–	1.1
Picric acid reagent	1.0	1.0	–
Distilled water	0.1	–	–
Creatinine standard	–	0.1	–
Buffer reagent	0.1	0.1	0.1

4. Set the zero using blank sample by taking optical density (O.D.) at 520 nm.
5. Plot the curve using the obtained result, and calculate the quantity of test sample.

Observation:

O.D. of standard sample ______________

O.D. of test sample ______________.

Formula:

$$\text{Serum Creatinine(mg\%)} = \frac{\text{O.D.Test}}{\text{O.D.Standard}} \times \frac{\text{Concentration of standard (mg)}}{\text{Volume of Serum}} \times 100 \quad (1)$$

Viva Voce:

Q1. What is creatinine?

Q2. What is true creatinine?

Q3. What are the condition of high serum creatinine level?

Q4. What is the creatinine coefficient?

Q5. What is the difference between creatine and creatinine?

Q6. What is the normal range of serum creatinine?

Q7. What is the principle of Jaffe's method for the estimation of serum creatinine?

Q8. What is the significance of creatinine coefficient?

Q9. What is the reason of red color Jaffe's complex in serum creatinine test?

Chapter 47

Determination of Blood Sugar

Blood sugar/glucose level is the extent of glucose present in the blood of a human or animal. Glucose is the most vital carbohydrate in the body. In the fed state, the majority of circulating glucose comes from the diet; in the fasting state, gluconeogenesis and glycogenolysis maintain glucose concentrations. Normally the blood sugar level ranges between 80 and 120 mg/100 ml of blood. In mild diabetic conditions, it ranges between 140 and 300 mg/100 ml of blood, and in severe diabetic conditions, blood sugar level has been observed up to 1200 mg/100 ml of blood. The decreased sugar level was observed due to Addison's disease, insulin administration, hypopituitarism, and hypoglycemia.

Body tries to maintain a constant supply of glucose for your cells by maintaining a constant blood glucose concentration. The concentration of glucose in blood, expressed in mg/dl, is defined by the term glycemia. The value of blood sugar in humans generally ranges from 70 to 100 mg/dl. Blood sugar levels are regulated by the hormones insulin and glucagon which act antagonistically. These two hormones are secreted by the islet cells of the pancreas, and thus are referred to as pancreatic endocrine hormones. When the blood glucose levels are high, insulin hormone is secreted causing the liver to convert more glucose molecules into glycogen, and when the blood glucose levels are low, glucagon is secreted and acts on liver cells to promote the breakdown of glycogen to glucose and increases the blood glucose concentrations. Essentially blood glucose levels determine the time of secretion of these hormones.

The blood glucose level is easily changed under the influence of some external and internal factors such as body composition, age, physical activity, and sex. Diabetes is a disease related by the abnormal metabolism of blood sugar and defective insulin production. So blood sugar levels are an important parameter for the study of diabetes. The level of glucose circulating in the blood at a given time is called as blood glucose level. The blood glucose level varies at different time on various part of the day. Hypoglycemia is a possible side effect of diabetes medications in which blood glucose

Aakanchha Jain et al. *Basic Techniques in Biochemistry, Microbiology and Molecular Biology: Principles and Techniques*, Springer Protocols Handbooks, https://doi.org/10.1007/978-1-4939-9861-6_47,

level drops below 70 mg/dl. In people with diabetes, the body doesn't produce enough insulin or respond to insulin properly. The result is that sugar builds up in the blood stream, damaging the body's organs, blood vessels, and nerves. This condition in which there is too much sugar in the blood stream is called hyperglycemia.

Collection of Blood Sample: Collect the blood sample in the tube containing potassium oxalate as anticoagulant and NaF as antiglycolytic agent in 3:1. Mix the sample thoroughly.

1 Folin-Wu Method

Principle: Glucose reduces the cupric ions present in the alkaline copper reagent to cuprous ions, or the cupric sulfate is converted into cuprous oxide, which reduces the phosphomolybdic acid to phosphomolybdous acid, which is blue when optical density is measured at 420 nm. The reaction depends on degree of alkalinity, heating time, and extent of temperature.

Reagents:

1. Sugar solution (1%): prepare by dissolving dextrose (1 gm) in 100 ml of saturated solution of benzoic acid (as preservative)
2. Alkaline copper solution: dissolve 2% Na_2CO_3 in 0.1 N NaOH (solution A) and 5% $CuSO_4$ in 1% sodium potassium tartrate or Rochelle salt (solution B); then mix 50 ml of solution A with 1 ml of solution B.
3. Phosphomolybdic acid: add 5 gm of sodium tungstate to 35 gm of molybdic acid; then add 200 ml of 10% NaOH and 200 ml of distilled water. Boil this solution for 20–30 min vigorously; this removes the ammonium present in molybdic acid. Then cool the solution, and dilute to about 350 ml, and finally add 125 ml of 85% orthophosphoric acid, and make the volume up to 500 ml with distilled water.
4. 10% sodium tungstate
5. Sulfuric acid (2/3 N)
6. Phosphoric acid solution (10%)
7. Sodium hydroxide solution (10%)

Procedure: 0.5 ml of blood is transferred to boiling tube containing 3.5 ml of water; then 0.5 ml of 10% sodium tungstate is added and mixed well followed by 0.5 ml of 2/3 N H_2SO_4 with shaking. It is allowed to stand for 10 min; it is then filtered using Whatman filter paper (paper no. 1). This filtrate is called tungstic acid blood filtrate and is taken as test sample. Transfer the 2 ml of blood filtrate to Folin-Wu tube graduated at 25 ml mark, and to another Folin-Wu tubes, transfer 2 ml of standard glucose solution and 2 ml of water as blank. Add alkaline $CuSO_4$ (2 mL) to each tree tubes so that the mixture is reached to the 4 ml mark of the Folin-Wu tube. Place the Folin-Wu tubes in boiling water bath precisely for 8 min. Then cool the tube under running water. Add phosphomolybdic acid (2 mL) to each tube. After 1 min it is diluted with water up to the mark. Then read the O.D. at 420 nm.

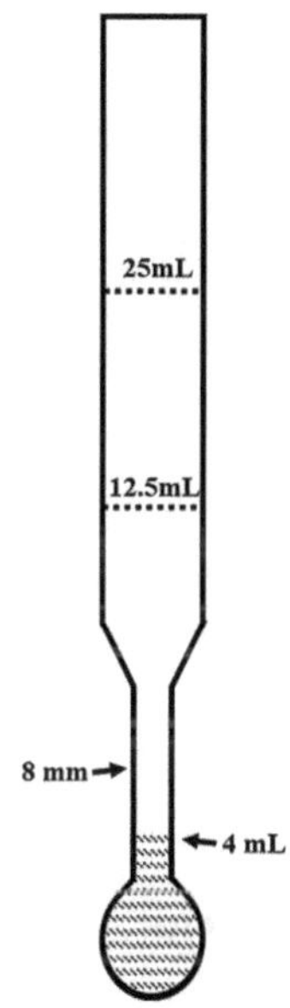

Normal Values:

- The normal blood sugar level ranges from 8 to 120 mg/100 ml of blood.
- In mild diabetic conditions, value of blood glucose from 140 to 300 mg/100 ml and in severe diabetic conditions value of up to 1200 mg/100 ml of blood have been noted.
- Low blood sugar level values are formed in insulin administration, Addison's disease, hypoglycemia and hypopituitarism.

Result:

_______ mg of glucose is present in 100 ml of given blood sample.

Calculation:

Mg of glucose/100ml of blood

$$= \text{mg of glucose in standard} \times \frac{\text{OD of test}}{\text{OD of standard}} \times \frac{100}{0.2} \quad (1)$$

2 Glucose Oxidase Method

Principle: Glucose oxidizes to gluconic acid in the presence of glucose oxidase. This further forms hydrogen peroxide. It is then broken down to oxygen and water by the enzyme peroxidase and then converted to a colored compound O-toluidine on reacting with oxygen.

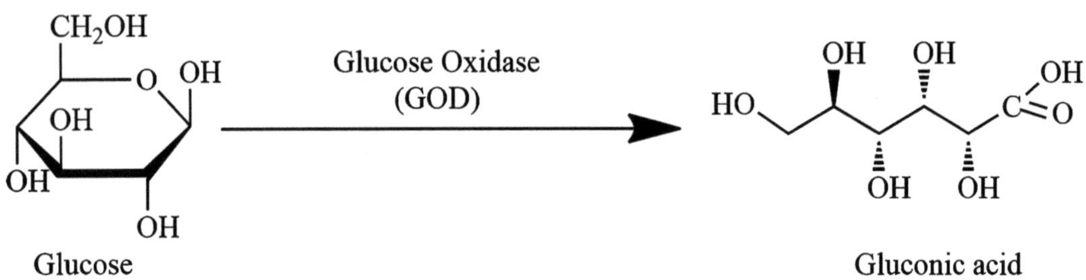

$$H_2O_2 \xrightarrow[\text{(POD)}]{\text{Peroxidase}} H_2O + 1/2\ O_2$$

Reagent:

1. Sodium hydroxide (0.3 N)
2. Zinc sulfate solution (5%)
3. Acetate buffer (0.15 M; pH -5)
4. O-toluidine (1%, in ethanol)
5. NaCl (0.9%)
6. Glucose oxidase reagent: Prepare this reagent fresh by dissolving 25 mg of glucose oxidase and 1% O-toluidine in the acetate buffer. Add a 5 mL of peroxidase and make up the volume up to 250 ml using buffer. Store this solution in a brown color bottle in refrigerator.
7. To 100 mL of benzoic acid solution dissolve 1 mg glucose. Dilute using acid to obtain solution of different concentrations (2.5, 5.0, 7.5, 10 mg/100 mL) which are equivalent to 50 mg/dl, 100 mg/dl, 150 mg/dl, and 200 mg/dl.

Procedure: For the test sample, pipette 0.1 ml of blood into 1.8 ml of sodium sulfate-zinc sulfate reagent in a centrifuge tube. Add 0.1 ml of 2 N sodium hydroxide, centrifuge at 3000 rpm for 5 min, and take 0.5 ml of supernatant in duplicate. Take 0.5 ml of distilled water as blank sample. Prepare standard concentration of glucose (200 mg/dl); use 0.5 ml of a range of glucose solutions (50 mg/dl, 100 mg/dl, 150 mg/dl, and 200 mg/dl) suitably diluted from standard. Add 5 ml of the glucose oxidase reagent incubate for 1 h at 37 °C and read the extinction at 540 nm against the reagent blank. If the absorbance reading of the sample is too high, dilute the supernatant which was obtained earlier, 2 times with distilled water and repeat the subsequent step.

Calculation:

$$\text{mg of glucose}/100\ \text{ml of blood} = \frac{\text{OD of test}}{\text{OD of standard}} \times 200\ \text{mg} \quad (2)$$

Viva Voce:

Q1. What is diabetes?

Q2. In Folin-Wu method, what are the reducing substance other than glucose?

Q3. What is the range of blood sugar by enzymatic method?

Q4. What is the range of blood sugar by Folin-Wu method?

Q5. What is the significance of shape of Folin-Wu tube?

Q6. What is true sugar?

Q7. Give the example of anticoagulant used in blood sugar estimation.

Q8. What is the principle of glucose oxidase method?

Q9. What is the key role of glucose oxidase in glucose oxidase method?

Q10. Name the reagents and their roles used in Folin-Wu method.

Chapter 48

To Perform Oral Glucose Tolerance Test

Oral glucose tolerance tests (OGTT) are used to measure how well the body can process a larger amount of sugar. If the blood sugar measured in the test is above a certain level, this could be a sign that sugar is not being absorbed enough by the body's cells. Diabetes or gestational diabetes might be at the root of this problem. In gestational diabetes, blood sugar levels are often higher due to changes in the metabolism during pregnancy—but they usually come back down again after the child is born. The glucose is most often given orally, so the common test is technically an oral glucose tolerance test (OGTT).

Cautions: Patients are advised to "avoid taking carbohydrate days or weeks before the test and should not have meal before going for the test." This also applies to all drinks except water. The test cannot be performed in illness because the results may reflect the patient's glucose metabolism whether healthy.

Procedure: First of all, the blood is drawn from the fingertip or vein or earlobe to determine the baseline of blood sugar level. Then, the patient is advised to drink given glucose solution (concentrated). For this 75 g, glucose is dissolved in 250–300 mL of water as it is recommended by WHO for oral dose in all adults. The amount to be given for children is adjusted by their body weight. If the test is being done to confirm suspected diabetes, blood is drawn again after 2 h, and the blood sugar level is measured. When testing for gestational diabetes, blood is drawn twice—first after 1 h and then again after another 2 h. If renal glycosuria is suspected, urine samples may also be collected along with fasting and 2 h blood test.

Aakanchha Jain et al. *Basic Techniques in Biochemistry, Microbiology and Molecular Biology: Principles and Techniques*, Springer Protocols Handbooks, https://doi.org/10.1007/978-1-4939-9861-6_48,

Table 1
Interpretatition of OGTT (WHO Diabetes criteria, 1999)

Glucose level	Normal		Impaired fasting glycemia IFG		Impaired glucose tolerance IGT		Diabetes mellitus DM	
Venous plasma	Fasting	2 h	Fasting	2 h	Fasting	2 h	Fasting	2 h
Mmol/l	<6.1	<7.8	≥6.1 and < 7.0	<7.8	<7.0	≥7.8	≥7.0	≥10
Mg/dl	<100	<140	≥100 and < 126	<140	<126	≥140	≥126	≥200

Interpretation of OGTT Results (Table 1):

- Fasting plasma glucose level should be below 6.1 mmol/L (100 mg/dl).
- Fasting level between 6.1 and 7.0 mmol/L (100–125 mg/dl) is borderline: impaired fasting glycemia.
- Fasting level at or above 7.0 mmol/l (126 mg/dl): diagnostic of diabetes.
- Glucose level after 2 h should be below 7.8 mmol/l (140 mg/dl).
- Level between 7.8 mmol/l (140 mg/dl) and 11.1 mmol/l (200 mg/dl): impaired glucose tolerance.
- Glucose level above 11.1 mmol/l (200 mg/dl): diagnosis of diabetes.

Viva Voce:

Q1. What is OGTT?

Q2. What is the fasting blood glucose range in normal health?

Q3. What is the range of glucose in fasting condition in diabetics?

Q4. What is the range of glucose in impaired glucose tolerance?

Q5. What is the need to perform OGTT?

Q6. What are the other health conditions that can arise due to high blood sugar levels?

Q7. How is OGTT done?

Q8. How much quantity is advisable according to WHO to perform OGTT?

Chapter 49

Determination of Serum Total Cholesterol

Cholesterol is a type of fat (lipid) present in the body that can be measured in blood serum and blood plasma and exists as free and esterified form. Cholesterol is vital constituent of the cell membranes, bile, brain, and nerve cells and benefits the body to absorb fats and fat-soluble vitamins. Cholesterol is also used in the formation of hormones such as cortisol, forms testosterone by the testicles, and forms estrogen and progesterone by the ovaries too. Normally it ranges from 150 to 200 mg/100 mL in adults. The level of cholesterol increases by 20–25% during pregnancy. The level of cholesterol increases in case of biliary obstruction, biliary cirrhosis, diabetes mellitus, hypothyroidism, nephrotic syndrome, and pancreatic disease.

1 Enzymatic Colorimetric Method

Principle: The hydrolysis of cholesterol esters occurs in the presence of cholesterol esterase. 4-Cholesten-3-one and H_2O_2 are then formed from the released free cholesterol by cholesterol oxidase. In the presence of phenol and peroxide, red quinoneimine derivative is formed from hydrogen peroxide (H_2O_2) and 4-aminoantipyrine that has an absorbance at 500 nm.

Reagents and Apparatus:

- Cholesterol buffer solution (cholesterol kit, 100 mL)—phenol 132 mg, NaH_2PO_4 0.78 g, Na_2HPO_4 0.71 g.
- Enzyme reagent (cholesterol kit, 100 mL dilution)—cholesterol oxidase 12 unit, cholesterol esterase 3.5 unit, peroxidase 6700 unit, 4-aminoantipyrine 17.0 mg/dL.
- Cholesterol standard solution (cholesterol kit)—300 mg/dL.
- Human serum sample, DI water bath.

Aakanchha Jain et al. *Basic Techniques in Biochemistry, Microbiology and Molecular Biology: Principles and Techniques*, Springer Protocols Handbooks, https://doi.org/10.1007/978-1-4939-9861-6_49,

Table 1
Sample setup of experiment

	Blank	Standard	Sample
Enzyme sol	3	3	3
Water	0.02		
Cholesterol		0.02	
Serum sample			0.02

Procedure: Prepare an enzyme solution by dissolving the enzyme reagent to 100 mL in the cholesterol buffer solution. Prepare the mixture as shown in Table 1. Place in a water bath at 37 °C for 5 min. Measure the absorbance of blank solution, standard solution and sample solution.

2 Sackett's Method

Principle: Serum cholesterol is extracted by alcohol-ether mixture. Cholesterol dissolved in chloroform after the evaporation of extract. Cholesterol gives green color when reacted with acetic anhydride and sulfuric acid.

Reagents: Chloroform, ether, and rectified spirit (alcohol).

Sample Preparation: To prepare stock standard solution of cholesterol, take 50 mg of cholesterol and dissolve in chloroform, and make volume up to 25 ml using chloroform. Dilute 1 ml of standard cholesterol solution up to 25 mL using chloroform.

Procedure: In a centrifugation tube, take mixture of ethanol-ether (10 mL), and to this mixture, add serum (0.2 mL). Shake the mixture gently. Keep the tube horizontally for 30 min to uniformly distribute the precipitates. Then centrifuge the tube and completely draw off the supernatant. Then dry the obtained residues and dissolve in chloroform (5 mL). Take standard cholesterol solution (5 mL) in another test tube, and add mixture of sulfuric acid and acetic anhydride (2 mL) in every tube. Mix the mixture of each tube well and keep at 25 °C for 15 min in the dark. Observe the reading using red filter at 660 Mμ. 7 mL of chloroform is considered as blank.

Calculation:

$$\text{Cholestrol (mg)} = \frac{\text{observation of unknown}}{\text{observation of standard}} \times 200 \quad (1)$$

3 Zak's Method

Principle: Serum when reacted with ferric chloride and acetic acid reagent precipitates the serum proteins. The cholesterol remains in supernatant undergo dehydration by ferric chloride; acetic acid and sulfuric acid produce red color compound. The strength of the color is proportional to the concentration of serum cholesterol content in the serum.

Reagent: Analytical grade concentrated sulfuric acid, standard cholesterol (0.04 mg/mL), and 0.05% ferric chloride solution in acetic acid.

Procedure: Into the series of test tubes, take different volume of cholesterol (standard) and take one tube as blank. Make up the volume up to 0.2 ml using water. Then to each tube, add $FeCl_3$-CH_3COOH reagent (3 ml). Mix thoroughly and centrifuge the test sample tube and take the 3 ml of supernatant in each tube. Add concentrated H_2SO_4 (2 ml). Mix the solution and heat on water bath for 10 min at 60 °C. Cool the mix at room temperature and take the readings at 540 nm. Calculate the concentration of test sample from standard curve of cholesterol.

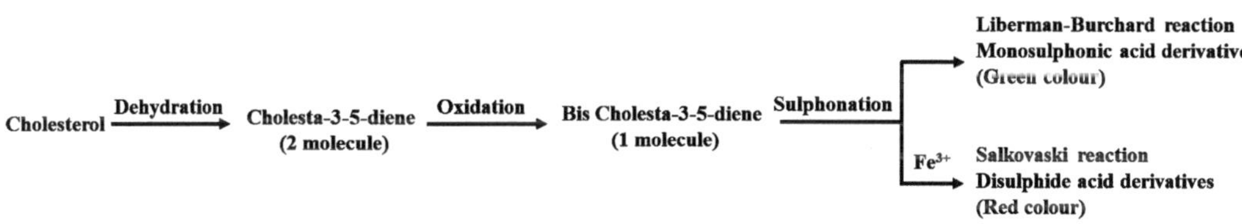

Calculation:

$$\text{Cholesterol (mg\%)} = \frac{\text{O.D. of test}}{\text{O.D. of Standard}} \times \frac{\text{Concentration of standard (mg)}}{\text{sample volume}} \times 100 \quad (2)$$

Viva Voce:

Q1. What is cholesterol?

Q2. What is the principle of Zak's method?

Q3. What is the normal range of cholesterol in adult?

Q4. What are the main causes of hypercholesterolemia?

Q5. What is the principle of enzymatic colorimetric method?

Q6. What are the major causes of increase in cholesterol level?

Q7. What are the symptoms of high serum cholesterol?

Q8. How do the statins lower the cholesterol?

Q9. How is serum cholesterol extracted?

Q10. Which product forms after oxidation of cholesterol?

Chapter 50

Preparation of Buffer Solution and Measurement of pH

The primary purpose of a buffer is to control the pH of the solution. Buffers can also play secondary roles in a system, such as controlling ionic strength or solvating species, perhaps even affecting protein or nucleic acid structure or activity. Buffers are used to stabilize nucleic acids, nucleic acid-protein complexes, proteins, and biochemical reactions (whose products might be used in subsequent biochemical reactions). To control the pH and to establish pH gradient, complex buffer systems are used in electrophoretic systems. Buffer solutions are composed of weak acids and bases that make them comparatively resistant to pH change. Theoretically buffers offer a ready source of both acid and base to either provide additional H+ if process consumes H+ or combine with excess H+ if a reaction generates acid.

Reagents:

- Acetic acid 0.2 M: glacial acetic acid 1.5 ml is made up to 100 ml by using distilled water.
- Citric acid: citric acid 2.10 gm in 100 ml distilled water.
- Dibasic sodium phosphate: 5.3 gm of disodium hydrogen phosphate in 100 ml distilled water.
- Diethyl barbituric acid.
- Monobasic sodium phosphate: 2.78 gm sodium dihydrogen phosphate in 100 ml distilled water.
- Sodium acetate solution: 0.64 gm of sodium acetate in 100 ml distilled water.
- Sodium bicarbonate solution: sodium bicarbonate 1.68 gm in 100 ml distilled water.

Aakanchha Jain et al. *Basic Techniques in Biochemistry, Microbiology and Molecular Biology: Principles and Techniques*, Springer Protocols Handbooks, https://doi.org/10.1007/978-1-4939-9861-6_50,

- Sodium carbonate solution 0.2 M:2.12 gm anhydrous sodium carbonate in 100 ml distilled water.
- Sodium citrate solution 0.1 M: sodium citrate 2.94 gm in 100 ml distilled water.
- Sodium diethyl barbiturate.

Procedure:

Acetic acid-sodium acetate buffer

Take a 100 ml flask, and to it add 36.2 ml of sodium acetate solution through a pipette, and then to it, add 14.8 ml glacial acetic acid. With the help of distilled water, make the total volume 100 ml. The resultant is 0.2 M of acetic acid-sodium acetate buffer. The pH is measured with the help of a pH meter. The electrode is washed with distilled water, and then it is dipped in the prepared buffer solution. The resultant pH is 4.6.

Barbitone buffer

Mix 2.85 gm of diethyl barbituric acid and 14.2 gm of sodium diethyl barbititrate in distilled water and then make the final volume 1000 ml. This is barbitone buffer. The pH is measured with the help of pH meter, and the final pH comes out as 6.8.

Citrate buffer

Mix 46.5 ml of citric acid with 3.5 ml of sodium citrate solution and make the final volume 100 ml with the help of distilled water. This is 0.1 M citrate buffer. The pH of this buffer is measured with the help of a pH meter, and the pH is 2.5.

Carbonate-Bicarbonate buffer

In a flask take 27.5 ml of sodium carbonate solution, and to it add 22.5 ml sodium bicarbonate solution. Then make the total volume 100 ml with the help of distilled water. This is 0.2 M carbonate-bicarbonate buffer. The pH meter is standardized, and the pH of the prepared buffer solution is measured. The pH is 10.2.

Phosphate buffer

Dihydrogen sodium phosphate (39 ml) is mixed with disodium hydrogen phosphate (61 ml), and the final volume is made to 200 ml with distilled water. This solution results in 0.2 M of phosphate buffer. The pH of phosphate buffer is measured with the help of pH meter, and it comes out 6.8.

Viva Voce:

Q1. What are buffer solutions?

Q2. What is the selection criteria of buffer?

Q3. How to adjust the pH of buffers?

Q4. What is the composition of carbonate-bicarbonate buffer?

Q5. What is the composition of barbitone buffer?

Q6. What is the composition of phosphate buffer?

Q7. On increasing the pKa of acids, what will be the effect on acids?

Q8. If a solution has to be a buffer, what should its pH be?

Q9. What is the role of disodium hydrogen phosphate in phosphate buffer?

Q10. What is buffer capacity?

Chapter 51

Study of Enzymatic Hydrolysis of Starch

Enzymes are protein in nature and catalyze chemical reaction in biochemistry. Enzymes are responsible for speeding up reaction, mostly synthesized in living cells. A study of enzymatic hydrolysis of starch will give knowledge about specific reactions of enzymes. There are several factors like temperature and pH that affect the reaction.

Reagents: Starch solution (1%): Take 10 gm of starch and dissolve it in 800 mL of distilled water, and then heat the solution to solubilize the starch. Make the volume up to 1000 mL with the same.

Phosphate buffer solution (0.1 M; pH 6.8).

Collect the salivary amylase by rinsing the mouth with distilled water and dispose of water.

Then collect approximately 5 mL of saliva in a test tube.

Iodine solution (1%).

Procedure: Take four test tubes, and to it add 1 ml of 1% starch solution and 1 mL of phosphate buffer solution. In a water bath, place all test tubes at temperatures 0 °C, 25 °C, 37 °C, and 95 °C. Then leave the test tube for few minutes, and then add about 1 ml of diluted saliva in each and every test tube. Now take four separate watch glasses, and to it add one drop of iodine solution. In this iodine solution, take one drop from each test tube, and add to the iodine to check if any starch is left or all hydrolyzed. If any starch is left, then it turns the iodine solution to blue color. Measure the time required for hydrolysis of starch in all test tubes.

Stages of hydrolysis by iodine test are shown in Fig. 1.

Aakanchha Jain et al. *Basic Techniques in Biochemistry, Microbiology and Molecular Biology: Principles and Techniques*, Springer Protocols Handbooks, https://doi.org/10.1007/978-1-4939-9861-6_51,

Product	Iodine Test
Starch	**Blue Colour**
Amylodextrin	**Purple Colour**
Erythrodextrin	**Red Colour**
Achrodextrin	**Colourless**
Maltose	**Colourless**

Fig. 1 Chamber setup for paper chromatography

Viva Voce:

Q1. Give the range of pH at which enzyme salivary amylase is most active.

Q2. Which of the following digestive enzyme is found in the saliva?

Q3. Name the glands from where the saliva is produced.

Q4. What is the old name for the enzyme salivary amylase?

Q5. Why does starch give a blue color with iodine?

Q6. Why does amylodextrin give a purple color with iodine?

Q7. What is the effect of temperature on iodine test?

Q8. What is the exact procedure for collection of saliva?

Q9. Enzymes are protein: true or false?

Q10. How does the pH affect the iodine test?

Chapter 52

Determination of Salivary Amylase Activity

Salivary amylase, formerly known as ptyalinis, a glucose-polymer cleavage enzyme that is produced by the salivary glands, breaks down starch into maltose and isomaltose. Amylase, like other enzymes, works as a catalyst. All catalysts are enzymes, but not all enzymes are catalysts. It comprises a small portion of the total amylase excreted, which is mostly made by the pancreas. Amylases digest starch into smaller molecules, ultimately yielding maltose, which in turn is cleaved into two glucose molecules by maltase. Starch comprises a significant portion of the typical human diet for most nationalities. Given that salivary amylase is such a small portion of total amylase, it is unclear why it exists and whether it conveys an evolutionary advantage when ingesting starch.

Principle: Amylase is the hydrolytic enzyme which breaks down many polysaccharides like starch, amylose, and dextrins and yields a disaccharide, i.e., maltose.

$$(C_6H_{10}O_5)_n + H_2O \rightarrow n(C_{12}\, H_{22}\, O_{11}) \qquad (1)$$

Reagents:

- Starch: Mix 1 g of soluble starch in 200 ml of 0.1 M phosphate buffer (pH 6.8), boil for 3 min, cool to room temperature, and filter if necessary.
- Enzyme: Saliva is the best and easily available source of amylase. Collect some saliva in a beaker and dilute it to 1:20 dilution with distilled water.
- 1% Sodium chloride: It is necessary for enzyme activity.
- Dinitrosalicylic acid (DNS): Dissolve NaOH(1.6 g) in 20 ml of distilled water. Take 1 g of 3,5 DNS in NaOH solution. In another beaker take 30 g of sodium potassium tartrate. Dissolve in

Aakanchha Jain et al. *Basic Techniques in Biochemistry, Microbiology and Molecular Biology: Principles and Techniques*, Springer Protocols Handbooks, https://doi.org/10.1007/978-1-4939-9861-6_52,

50 ml of distilled water. Mix this DNS solution and finally make the volume up to 100 ml with distilled water.

- Standard solution of maltose: It is prepared by dissolving 200 mg maltose in 100 ml of water.

Procedure: Take 0.5 ml of substrate and 0.2 ml of 1% NaCl in a test tube and preincubate at 37 °C for 10 min, and then add 0.3 ml of diluted saliva and incubate for 15 min at 37 °C. Stop the reaction by addition of 1 ml of DNS reagent, mix well, and keep the test tubes in boiling water bath for 10 min. Cool and dilute with 10 ml of distilled water. Read the color developed at 520 nm. Simultaneously set up the color developed at 520 nm. Simultaneously set up the blank as per the test by adding DNS prior to the addition of enzyme simultaneously. Set up the standards of different test tubes and repeat the experiment as per the test and measure the color developed at 520 nm absorbance (Table 1).

Preparation of Phosphate Buffer: Dissolve 0.2 M (2.7218 g) of KH_2PO_4 in 100 ml of distilled water to this solution, and add 0.5 M (2.8053 g) KOH drop by drop till the pH is set to 6.8. Then make it to 200 ml with distilled water. So the final concentration is 0.1 M of 200 ml phosphate buffer.

Result: The amount of maltose in the given unknown sample is ________ grams of maltose formed per 100 ml of enzyme per 1 h.

Table 1
Absorbance/OD (Optical Density) of sample at 520 nm

S. no.	Volume of standard (ml)	Volume of distilled water (ml)	Concentration of standard (ml)	Volume of (ml)		Volume of distilled water (ml)	OD at 520 nm
1	Blank	1.0	0.0	1.0	Boil water bath for 10 minutes, and cool	10	
2	0.2	0.8	0.4	1.0		10	
3	0.4	0.6	0.8	1.0		10	
4	0.6	0.4	1.2	1.0		10	
5	0.8	0.2	1.6	1.0		10	
6	1.0	0.0	2.0	1.0		10	

Viva Voce:

Q1. What are the end products of salivary amylase activity?

Q2. How many pairs of salivary glands are found in human body?

Q3. Why is NaCl solution added in the starch solution while testing salivary amylase activity?

Q4. The starch present in experimental tubes has been completely digested. How will you confirm the above observation in this experiment?

Q5. What is the composition of human saliva?

Q6. How is the presence of starch tested in the reaction mixture?

Q7. How will you select the wavelength for color identification?

Q8. How many pairs of salivary glands are found in human beings?

Q9. What are the end products of salivary amylase activity?

Q10. Why is NaCl added in the starch solution while testing salivary amylase activity?

Chapter 53

Study the Effect of Temperature on Salivary Amylase Activity

Enzymes are proteinaceous in nature and catalyze chemical reaction in biochemistry. Enzymes are responsible for speeding up reaction and mostly synthesized in living cells. A study of enzymatic hydrolysis of starch will give knowledge about specific reactions of enzymes. There are several factors like temperature and pH that affect the reaction. At higher temperature the enzymes are denatured, while at lower temperature, the enzymes are deactivated, so this takes more time at low and high temperature to digest the starch. At optimum temperature (32–37 °C), the enzyme is active and therefore consumes less time for starch digestion.

Reagents:

1. Starch solution (1%): Take 10 g of starch and dissolve it in 800 mL of distilled water, and then heat the solution to solubilize the starch. Make the volume up to 1000 mL with the same.
2. Preparation of phosphate buffer: Dissolve KH_2PO_4 (2.7218 g) of in 100 ml of distilled water. To this solution add KOH (2.8053 g) dropwise till the pH is set to 6.8. Make the volume up to 200 ml using distilled water. So the final concentration is 0.1 M of 200 ml phosphate buffer.
3. Enzyme source (saliva): Collect the salivary amylase by rinsing the mouth with distilled water and dispose of water. Then collect approximately 5 mL of saliva in a test tube.
4. 1% NaCl.

Aakanchha Jain et al. *Basic Techniques in Biochemistry, Microbiology and Molecular Biology: Principles and Techniques*, Springer Protocols Handbooks, https://doi.org/10.1007/978-1-4939-9861-6_53,

Table 1
Observation of color of the iodine at different time and temperature

Test tube	Temperature	Contents	Observation (seconds)			
			5	15	45	60
Control	37 °C	Starch + saliva + iodine solution				
Test	5 °C	Starch + saliva (cooled) + iodine solution				
	50 °C	Starch + saliva (boiled) + iodine solution				

Procedure: In a beaker take 5 ml of starch solution and 1 ml NaCl solution. Take three test tubes and pour this solution and mark them as A, B, and C. Maintain the temperature of the beaker containing ice cubes at 5 °C. Transfer the tube A into a beaker containing ice. Keep the second experimental tube B into water bath at 37 °C and third tube C into the beaker and maintain the temperature at 50 °C. Using a dropper, take 1 ml saliva solution using dropper in test tube A. Likewise, add 1 ml saliva solution into test tubes B and C and test tube C. Instantly, take few drops from tube A and transfer this into test tubes containing iodine solution. Do the same procedure for test tubes B and C. Note this time as zero-minute reading. Repeat the experiment at an interval of every 2 min till color of iodine does not change (Table 1).

Viva Voce:

Q1. What is meant by optimum temperature for enzyme activity?

Q2. What is the optimum temperature for the enzymatic activity of salivary amylase?

Q3. Why enzymes get denatured at high temperature or deactivated at low temperature and in the presence of alcohol or formaldehyde?

Q4. How many pairs of salivary glands are found in human beings?

Q5. What are the end products of salivary amylase activity?

Q6. Why is NaCl added in the starch solution while testing salivary amylase activity?

Q7. How will you prepare standard starch solution?

Q8. The starch present in experimental tubes has been completely digested. How will you confirm the above observation in this experiment?

Q9. What is the composition of human saliva?

Q10. How is the presence of starch tested in the reaction mixture?

Chapter 54

Study the Effect of Substrate Concentration on Salivary Amylase Activity

It has been rumored in the theory of enzyme action that the enzyme forms a combination with the substrate which breaks up to yield the products of the reaction and the free enzyme again. Further assumptions, such as that the velocities of combination of enzyme and substrate and of the decomposition of their compound, follow the law of mass action and permit the deduction of mathematical expressions which connect the velocity of the reaction as a whole, at least in the initial stages, with the concentration of enzyme and substrate.

Principle: With a fixed concentration of enzyme, an increase of substrate will result at first in a very rapid rise in velocity or reaction rate. As the substrate concentration continues to increase, however the increase in rate of reaction begins to slow down until, with a large substrate concentration, no further change in velocity is observed. Michealis and others reasoned correctly that an enzyme-catalyzed reaction at varying substrate concentrations is diphasic, i.e., at low substrate concentration, the active sites on molecules (enzyme) are not saturated by substrate, and the enzyme rate varies with substrate molecules concentration. As the number of substrate molecules increases, the sites are covered to a greater degree until at saturation no more rites are available, the enzyme is working at full capacity, and now the rate is independent of substrate concentration.

Reagents: Starch solution and stock enzyme solution, iodine.

Aakanchha Jain et al. *Basic Techniques in Biochemistry, Microbiology and Molecular Biology: Principles and Techniques*,
Springer Protocols Handbooks, https://doi.org/10.1007/978-1-4939-9861-6_54,

Table 1
Observation of different concentration and substrate on enzyme activity

S.NO.	Substrate (ml)	Enzyme (ml)	Time (min)	DNS (ml)		H_2O (ml)	OD (540nm)	Maltose (mg)	Activity (mol/ml/min)
1					Heat in boiling water bath for 5min and cool				
2									
3									
4									
5									
6									
7									

Procedure: Prepare a test tube rack containing 4 rows of 10 colorimeter tubes each. Label the first tube in each row 1:2, 1:4, 1:8, and 1:16, corresponding to their respective dilutions of starch. Now take Erlenmeyer flask and label it depending on the respective dilutions, and then to each flask, add 50 ml of distilled water. Now to each flask, add 50 ml of starch solution. Mix it properly, and from this, transfer 50 ml of the mixed solution to the second flask and then again mix it properly. Repeat the same with the third and fourth flask. Then from fourth flask, discard 50 ml of the solution. Now take 1 ml pipette and transfer 0.2 ml of the enzyme solution to all flask. Mix it properly and transfer 3 ml of the solution from each flask to colorimeter tube. To each tube add 3 drops of iodine and allow it to mix properly. Then place the tubes in the colorimeter for measuring absorbance at 540 nm. Calculate the transmittance percent. Do this for every flask with 2 min gap.

Result: The tubes with the most substrate and fewer enzymes were darker in color and took longer to change to iodine color. The tubes with the lower concentration of substrate and the high concentration of enzymes turned iodine color much faster (Table 1).

Conclusion: With an increase or decrease in substrate concentration, the % transmittance and absorbance will increase or decrease.

Viva Voce:

Q1. What is the effect of concentration of substrate on enzymatic activity?

Q2. Name the parameters which affect the enzymatic activity.

Q3. What is Km in Michaelis-Menten equation?

Q4. What is the effect of substrate on % transmittance and absorbance?

Q5. What is the composition of human saliva?

Q6. How is the presence of starch tested in the reaction mixture?

Q7. How many pairs of salivary glands are found in human body?

Q8. In enzyme assays, what is kept constant except?

Q9. Which compound forms on action of salivary amylase on starch?

Q10. Name the example of ion which activates salivary amylase activity.

Chapter 55

Analysis of Butter

Butter is the most highly concentrated forms of fluid milk. Butter is the product made exclusively by churning the cream which has been separated from warm milk. Excess water is removed, and the mass rendered homogenous by addition of salt or without salt contains $\geq$80% milk fat by weight. Accordingly, reduction of fat or addition of oils or emulsifiers to enhance the textural features of butter raises issues related to standard of identity. Fractionation of milk fat seems to be the eminent substitute to produce modified milk fat that can be converted into butter of improved functional and nutritional value.

Properties: Butter consists of fat globules and moisture droplets embedded in a continuous phase of butterfat. Flavor of butter is due to diacetyl and to a lesser extent to butyric, acidic, propionic, and formic acid acetaldehyde acetoin. Diacetyl content in butter can be increased by addition of suitable culture.

Composition of Butter:

- Butterfat
- Water
- Curd (casein, lactose, and mineral matter)
- Salt
- Vitamin A
- Milk solids other than fat
- Preservatives
- Colorants
- Flavor

Aakanchha Jain et al. *Basic Techniques in Biochemistry, Microbiology and Molecular Biology: Principles and Techniques*, Springer Protocols Handbooks, https://doi.org/10.1007/978-1-4939-9861-6_55,

Analysis of Butter: The routine analysis of butter includes the following:

1. Determination of water
2. Determination of salt
3. Examination of preservatives

Filtered fat examined for:

1. Reichert value
2. Polenske value
3. Rancidity value

Sampling: The sample is placed in screw capped jar and warm in oven at 32–35 °C. Shake vigorously from time to time in order to produce a homogenous fluid emulsion.

1. **Water**: Weigh 3–4 gm of sample into metal dish, dry in a boiling water bath with stirring and then in an oven at 100 °C (approximately 3 h), and weigh again. The difference in the weight gives the amount of water present (The limit is not more than 16%).
2. **Curd and salt:** Melt the residue obtained above by gentle heating, add light petroleum (to melted fat), stir, and filter through tarred filter paper (The filtrate is used for fat determination by methods given below). Dry the paper in oven at 100 °C and weigh the curd and salt.
3. **Salt**: Transfer the curd and salt together with filter paper to a beaker containing hot water, cool and titrate with 0.05 M silver nitrate ($AgNO_3$) solution using potassium chromate as indicator.

Other Methods (Volhard's Method):

Mix 2 gm of sample butter with 10 ml of water and 25 ml of silver nitrate. Add 10 concentrated nitric acid and 1ml nitrobenzene to it. Titrate excess of silver nitrate with 0.05 M potassium thiocyanate using ferric alum as indicator. Calculation:

$$1 \text{ mL of } 0.05\text{M } AgNO_3 \equiv 0.0002925 \text{ gm NaCl}$$

Use: Salt in butter act as a preservative by retarding the growth of most bacteria and molds.

Limit: 0.03–1.8%

For salted butter: Not more than 3%

4. Curd: It is obtained by subtracting the salt and water content from sample. Curd consists of casein (protein), lactose, and lactic acid.

Proteins can be determined in curd plus salt residue by the Kjeldahl method using the factor N × 6.38.

Lactose can be determined on the curd by Lane and Eynon process. Butter usually contains between 0.18 and 0.40% lactose.

5. Milk solids: Milk solid other than fat can be calculate by

$$\text{Non} - \text{fat milk solids (NFMS) (\%)} \\ = 100\ (\text{Percentage of water} + \text{Fat} + \text{Salt})$$

The NFMS consists essentially of curd (protein), lactose, and lactic acid.

Limit: Not more than 2%

6. Ash: The cream used by butter manufactures is sometimes neutralized by sodium carbonate (or) bicarbonate which increases the ash value.
7. Titrable acidity: It is used to estimate the amount of lactic acid present.

   Briefly, take 90 mL of boiled water and 18 gm of butter, shake them, and titrate with 0.02 M NaOH using phenolphthalein as indicator.

$$\text{Titrable acidity} = \frac{0.02\text{M NaOH required (mL)}}{100} \tag{1}$$

   Limit: Not more than 35%

8. Determination of boric acid by turmeric test:

   Pour away the upper fatty acid layer by melting the 5–10 gm of butter. Then pour the aqueous liquid remaining into a flat porcelain dish, add three drops of saturated oxalic acid solution and two drops of 5 M hydrochloric acid and concentrated alcoholic extract of turmeric, and evaporate on water bath. In the presence of boric acid, a reddish residue is produced.
9. Added color: Divide an ethereal solution of isolated butterfat into two tubes. To one tube (A) is added 1–2 mL hydrochloric acid and to (B) 1–2 ml 10% sodium hydroxide solution. If azo dyes are present, tube A will show red color, but tube B will not show any color. If annatto or other vegetable color is present, there is no color in A, but a yellow color appears in B.

10. Trace metals: It has been shown that the presence of 1–2mg/kg of copper in butter encourages fat rancidity with production of fishy and tallowy taint in a few days. Rancidity may also be enhanced by other transition metals such as chromium, iron, manganese and nickel, but the action is less marked. The rancidity formation appears to increase with increasing acidity. For the milk products, BS 6394: Part 1 1983 (ISO 5738) describes a calorimetric method for copper using sodium diethyldithiocarbamate, and Part 2 (ISO 6732) describes a colorimetric method for iron using o-phenanthroline. Atomic absorption spectrophotometry has largely replaced colorimetric method.

Examination of Butterfat: The usual range of constants for genuine butterfat is given in Table 1

Butterfat in its triglycerides contains varying percentage of butyric, capric, caproic, caprylic, stearic, lauric, myristic, palmitic and oleic acid together with small quantities of other acids. Many are based on determination of relatively high proportion of butyric acid in butterfat using chromatography.

Preparation of sample:

The fat sample must be free from water and non-fatty materials. It can be obtained by filtration of melted fat as described in the Reichert process.

Table 1
Standard parameters of butterfat

Property	Max	Min	Mean
Specific gravity at 37.8 °C/37.8 °C	0.193	0.910	0.912
Refractive index at 40 °C	1.4561	1.4524	1.4548
RI on Zeiss butyro scale	45.5	40.0	43.5
Saponification value	232	222	228.5
Iodine value	40	26	36
Reichert value	32.8	24.5	30
Polenske value	3.5	1.5	2.8
Kirschner value	27	21	24
Baryta value	−0.7	−23.8	−9.6
Mean molecular weight of fatty acid	267	258	269

Physical parameters: **Density**: It can be determined by weighing the melted fat and then water in pycnometer at 37.8 °C. Due to high proportion of glycerides of low molecular weight, butterfat has a higher density than most other fats. The difference between the density of butterfat and other fats is almost at a maximum at 37.8 °C.

Chemical Analysis: Chemical tests are commonly used for accessing the rancidity of butterfat and proportions of other fats in butterfat.

Isolation of butterfat for analysis

Heat a portion of butterfat in a beaker to a temperature of 50–60 °C until the fat separates from water and curd. Filter the fat layer through a paper into dry vessel in an oven at a temperature above the solidifying point of fat. If necessary refilter the filtrate under the same conditions until it is clear and free from water. Liquefy the fat completely and mix before taking samples for analysis.

Determination of volatile acids in butterfat can be done by three main methods namely, Reichert, Polenske, and Kirschner values

(A) Definitions.

The Reichert value is the number of milliliter of 0.1 M aqueous alkali solution required to neutralize the water-soluble volatile fatty acids distilled from 5 g of fat under predefined conditions.

The Polenske value is the number of milliliters of 0.1 M aqueous alkali solution to neutralize the water-insoluble fatty acid distilled from 5 g of fat under predefined conditions.

The Kirschner value is the number of milliliters of 0.1 M aqueous alkali solution to neutralize the water-insoluble fatty acid which forms water-soluble silver salt distilled from 5 g of fat under predefined conditions

(B) Apparatus.

(a) Graduated cylinder 100 mL capacity.

(b) Pipette of 50 mL capacity.

(c) Distillation apparatus.

(C) Reagents.

(a) Glycerol.

(b) Sodium hydroxide, 50% (w/w) solution.

(c) Dilute sulfuric acid: 25 ml of concentrated H_2SO_4 is added to 1 liter with water and adjust until 20 ml neutralizes 1 ml of 50% NaOH solution.

(d) Pumice powder, ground passing through a BS 50 mesh test sieve and remaining on a BS 90 mesh test sieve.

(e) Phenolphthalein indicator (0.5% solution in 95% ethanol).

(D) Procedure.

Weight 5 ± 0.01 g of the fat prepared as described above into a Polenske flask. Add 20 g of glycerol and 2 ml of the 50% NaOH solution. Protect the burette containing the latter from the carbon dioxide and wipe its nozzle clean from carbonate deposit before withdrawing solution for test; reject the first few drops withdrawn from the burette. Heat the flask over the naked flame, with continuous mixing, until the fat is saponified, including any drops adhering to the upper part of the flask, and the liquid becomes perfectly clear, avoiding overheating during this saponification. Cover the flask with the watch glass.

Make a blank test without fat, but using the same quantities of reagent and following the same procedure, again avoiding overheating with sodium hydroxide, such overheating would be indicated by darkening of the solution.

Measure 93 ml of boiling water, vigorously boiled for 15 min before use into 100 ml of graduated cylinder. When the soap is sufficiently cool to permit addition of water, without loss, but before the soap has solidified, add the water, draining the cylinder for 5 s and dissolve the soap. If the solution is not clear or is darker (incomplete saponification) than light yellow (due to overheating), repeat the saponification with the fresh sample of the fat.

When the distillate reaches to 110 ml mark, remove the flame and replace the 110 ml flask by a cylinder of about 25 ml capacity to catch draining. Close the 110 ml flask with stopper and without mixing the contents place it in water bath at 15 °C for 10 min so as to immerse the 110 ml mark. Remove the flask from the water, dry the outside, and invert the flask carefully, avoiding the stopper with the insoluble acid. Mix the distillate by four or five double inversions, without violent shaking, filter through Whatman filter paper No. 4 which fits snugly into a funnel. Reject the first running and collect 100 ml in dry flask; cork the flask and retain the filtrate for titration as (R) at below.

The filtrates should be free from insoluble fatty acids. In cases where liquid insoluble fatty acids pass through the filter, receive the filtrate in a separating funnel, and after separation draw off the lower (aqueous) layer, into a dry 100 ml flask, leaving behind insoluble acids that have risen to the surface. Add these to the main bulk of the insoluble acids.

Detach the still head and wash the condenser with three successive 15 ml portions of cold water passing each washing separately through the cylinder, the 110 ml flask, the separating funnel if used, and the filter, nearly filling the paper each time and draining each washing before filtering the next. Discard the washings. Dissolve the insoluble acids by three similar washings of the condenser, the cylinder, and filter with 10 ml of neutralized ethanol, collecting the solution in the 110 ml flask and draining the ethanol after each washing. Cork the flask and retain the solution for titration, as at (P) below.

(R) Reichert, or Soluble Volatile Acid, Value: *Pour 100 ml of the filtrate containing the soluble volatile acids titration flask, add 0–1 ml of phenolphthalein indicator, and titrate with the barium hydroxide solution until the liquid becomes pink, rinsing the 100 ml flask with the nearly neutralized liquid toward the end of the titration (0–1 M may be used for the titration if the Kirschner value is not required). Kirschner value is to be obtained, and the titration flask must be dry before use. Note the actual volume of barium hydroxide solution used; drain the 100 ml flask into the titration flask and close with a cork and continue at (K) below.*

(P) Polenske or insoluble volatile acid value: *Titrate the alcoholic solution of the insoluble volatile acids after addition of 0.25 ml of phenolphthalein indicator with the 0.05 M barium hydroxide or 0.1 M sodium hydroxide solution until the solution becomes pink.*

(K) Kirschner Value: *Add 0.5 g of finely powdered silver sulfate to the neutralized solution from above. Allow the flask to stand in the dark for 1 h without shaking and filter the contents in the dark through a dry filter. Transfer 100 ml of the filtrate to a dry Polenske flask, and add 35 ml of cold distilled water, recently boiled for 15 min, 10 ml of the dilute sulfuric acid, and a loosely wound 5 mm of 30 cm of aluminum wire (about 1 mm thick or 18–20 SWG) or 0.1 g of pumice powder. Connect the flask with the standard apparatus and repeat the process but without the cooling for 10 min, filter and the titrate of 100 ml of the filtrate with the barium hydroxide solution.*

Calculation:

$$\text{Reichert Value} = 1.10(T_1 - T_2) \quad (2)$$

$$\text{Polenske Value} = (T_3 - T_4) \quad (3)$$

$$\text{Kirschner} = \frac{121(100 + T_1)(T_5 - T_6)}{10000} \quad (4)$$

where:

T_1 = ml of 0.5 M barium hydroxide solution used for sample under (R)

T_2 = ml of 0.5 M barium hydroxide solution used for blank under (R)

T_3 = ml of 0.5 M barium or 0.1 M sodium hydroxide solution used for sample under (P)

T_4 = ml of 0.5 M barium or 0.1 M sodium hydroxide solution used for blank under (P)

T_5 = ml of 0.5 M barium hydroxide solution used for sample under (K)

T_6 = ml of 0.5 M barium hydroxide solution used for blank under (K)

Viva Voce:

Q1. What is the composition of butter?

Q2. Give the name of parameter(s) for the routine analysis of butter.

Q3. What is rancidity value?

Q4. What is Polenske value

Q4. Define Reichert value.

Q5. How will you determine lactose in curd?

Q6. What is the effect of trace metals on rancidity?

Q7. Which apparatus is used for determination of density?

Q8. What is titratable acidity?

Q9. What is the reason for the flavor of butter?

Q10. What the nitrobenzene do in Volhard's method?

Chapter 56

Analysis of Milk

Composition of Milk: Cow's milk has been defined as the secretion, excluding colostrum, which can be gained by normal milking methods from the lactating mammary gland of the healthy, normally fed cow. Milk can be considered as containing three basic components, namely, water, fat, and non-fatty solids (NFS) or solids-not-fat (SNF). The organic matter in the non-fatty portion consists mainly of casein and whey proteins together with lactose and lactic and citric acids.

Milk Proteins: Two main forms of protein occur in milk, micellular proteins (75–85%), which are caseins associated with calcium, phosphate, and citrate, having an open, random structure, and which precipitate at pH 4.6, and serum or whey proteins (15–22%) differing in their molecular, physical, and functional properties but having globular structure and that are soluble at pH 4.6. There is also 2–4% of a proteose-peptone fraction.

The percentage composition of the casein fraction is α_{S1}-casein, 39–46; α_{S2}-casein, 8–11; β-casein, 25–35; κ-casein, 8–15; and Υ-casein, 3–7. The percentage composition of the whey protein fraction is β-lactoglobulin, 7–12; α-lactoglobulin, 2–5; serum albumin, 0.7–1 3; and immunoglobulins, 1.9–3.3 (Brunner 1981).

Milk protein fractions are important food ingredients, functional additives, and industrial chemicals (Morr 1982; Swartz and Wong 1985).

Acid casein is obtained by adjusting skim milk to pH 4.6 with mineral acid or by lactic fermentation. The precipitated curd is separated, washed, and dried and is low in calcium and phosphate, which remains in solution.

Rennet casein is obtained as a coagulate caused by the specific action of chymosin, a proteolytic enzyme in rennet. It retains the calcium-phosphate complex and can be solubilized by sequestering the calcium by the addition of sodium phosphate.

Caseinates are manufactured by solubilizing casein with alkalis or sequestering agents and drying the resultant product. Sodium

Aakanchha Jain et al. *Basic Techniques in Biochemistry, Microbiology and Molecular Biology: Principles and Techniques*, Springer Protocols Handbooks, https://doi.org/10.1007/978-1-4939-9861-6_56,

caseinates form translucent solutions of high viscosity and water-binding properties, whereas calcium caseinates form opaque solutions with low viscosity. They are ingredients of coffee whitener products.

Whey powder is prepared by drying the serum left after the separation of casein skim milk and contains 70–80% of lactose and 9% of protein. Whey protein concentrates are prepared by removing lactose and soluble ions from whey by ultrafiltration, to yield powders containing 35–80% of protein.

Lactalbumin is precipitated out when whey is heated to 90 °C and consists of insoluble-denatured whey proteins. Co-precipitates are formed by co-reacting casein and denatured whey protein. The precipitated material will have varied functional properties depending on the heat applied, pH, and calcium content.

Acidity and pH: Cow's milk is acid to phenolphthalein, alkaline to methyl orange, but amphoteric to litmus due to the presence of phosphates. The pH is usually between 6.4 and 6.6. The total acidity of freshly drawn milk is usually about 0.14% as lactic acid. On storage, the acidity increases due to the action of microorganisms, and a sour taste is perceptible when this reaches about 0.3%. Milk is decidedly sour at 0.4%, and when the acidity reaches 0.6%, it curdles at ordinary temperatures. A different spoilage pattern is set up in heat-treated milk due to the change in the microbial flora.

Mineral Matter: The total ash of genuine milk varies from about 0.71–0.75% and for normal milk constitutes about 8.3% of the non-fat solids.

Experiments on Chief Constituents:

1. Dilute 10 ml. of milk by 10 ml. of water. Then add 2% acetic acid, drop by drop till a heavy precipitate of casein is formed. Allow it to stand for 5 min and filter. Wash the residue with water and filter again. Repeat the washing thrice.
2. A small portion of residue is allowed to dissolve in 4 mL of 2% Na_2CO_3, and with this solution, perform (a) biuret test, (b) ninhydrin test, and (c) color test for amino acids to justify that residue contains protein.
3. Take a portion of the residue and perform other tests.
4. A portion of the dried residue is shaken up vigorously with ether in a test tube for about 1–2 min, and allow it to stand for a while. Decant the supernatant ether layer in a tiry basin and allow it to evaporate spontaneously at room temperature. Rub a piece of ordinary writing paper on the portion left in the basin. A translucent grease spot on the paper indicates the presence of fat.

5. To 10 ml of the filtrate, add a drop of chlorphenol red indicator. Then, add 2% Na_2CO_3 solution or 2% acetic acid to bring the pH 5.4 showing the change of red to yellow color. Boil the solution. A coagulum appears owing to the coagulation of albumin and globulin. Filter and keep the filtrate for the following experiments.
6. With the portion of this filtrate, perform Molisch's test to know the presence of carbohydrate and other tests for the detection of the nature of sugar present.
7. Perform silver nitrate test for detecting chloride with a portion of the filtrate.
8. To the remaining portion of the filtrate, add a few drops of NH_4OH to make it alkaline and to slight gelatinous precipitate magnesium phosphates that are produced. Filter dissolves the residue on the filter paper by pouring over 5 ml of hot 2% acetic acid. (a) To a portion of this solution add clear solution of potassium oxalate. A turbidity of calcium oxalate appears. (b) To the remaining solution, add 2 ml ammonium molybdate solution. Mix and heat. Lemon yellow precipitate of phosphomolybdate indicates the presence of phosphate.

Analysis of Milk: Methods and techniques for sampling milk and milk products for microbiological, chemical, physical, and sensory analysis are described in BS: 1985 (ISO 707-1985). In the routine examination of milk, it is convenient to take lactometer (density) reading, and after determining the fat by Gerber method, the total solids and non-fatty solids content can be calculated.

Total Solids: The following is an outline of the gravimetric method for determining the total solids of milk.

Weigh out about 5 g of mixed milk into a weighed flat-bottomed metal dish (about 7 cm diameter) and provide with a closely fitting lid, which is weighed with the dish. Place the uncovered dish on a boiling water either for 30 min. or until most of the moisture is driven off. Then wipe the bottom of the dish and transfer it to a well-ventilated oven at 102 ± 2 °C. Place the lid next to the dish in the oven. Dry for 2 h in the oven and then open the dish with the lid, cool for 30 min in a desiccator, and weigh. Then, heat the dish and lid as before for 1 h in the oven, cool, weigh, etc., until the loss of weight between successive weighing does not exceed 1 mg. If the acidity of milk exceeds 0.20% (as lactic acid), some of the acid present may volatalize on drying, and neutralizing of the milk with 0.05 M strontium hydroxide is suitable. If this is done, a deduction must be made from the weight if the residue obtained is equivalent to 0.00428 g for each ml of 0.05 M strontium hydroxide used.

Fat: Rapid volumetric methods are often used for routine purposes for determining fat in milk. In the UK, the Gerber method is commonly used: butyrometer.

The test is performed in butyrometer tubes that are closed with special metal and rubber stoppers. For the test, pipette the following liquids into a milk butyrometer, ensuring that they do not mix with one another: 10 ml of sulfuric acid (density 1.812–1.818 at 20 °C ≡ 90% m/m approx.), 10.94 ml of mixed milk and 1 ml fat-free amyl alcohol. These are centrifuged at 1100 r.p.m for 4 min. Transfer the tube (stopper downwards) to a water bath at 65 °C for at least 3 min and read off the percentage fat directly from the scale (extending from the water of upper meniscus to the flat bottom the fat column). Reading to an accuracy of 0.5% is usually adequate for routine purposes. As it is difficult to separate the small fat globules in homogenized (e.g., sterilised) milk, it is advisable to recentrifuge after warming in the 65 °C bath until the reading reaches a maximum.

Rose-Gottlieb Method for Fat Analysis in Milk: The gravimetric Rose-Gottlieb method is prescribed as the reference method for determining fat in milk.

Principle: The sample is treated with ammonia and ethyl alcohol, the former to dissolve the protein and the latter to help precipitate the proteins. Fat is extracted with diethyl ether and petroleum ether. Mixed ethers are evaporated and the residue weighed. This method is considered suitable for reference purposes. Strict adherence to details is essential in order to obtain reliable results.

Reagents:

1. Concentrated ammonium solution-specific gravity 0.8974 at 16 °C
2. Diethyl ether, peroxide-free-specific gravity 0.702
3. Ethyl alcohol (95–96% v/v)

It may be maintained free from peroxides by adding wet zinc foil (approximately 80 cm/L, cut in strips long enough to reach at least halfway of the container) that has been completely immersed in dilute acidified copper sulfate for 1 min and subsequently washed with water.

1. Petroleum ether, boiling range 40–60 °C
2. Mixed solvent (Prepared by mixing equal volumes of the ether and light petroleum.)

Procedure: Weigh accurately about 10 g of sample (liquid milk) and transfer to extraction tube. Add 1.0 ml of ammonia specific gravity 0.8974; mix well in the lower bulb thoroughly. Add 10 ml ethyl alcohol and mix by allowing the liquid to flow backwards and forwards between the two bulbs. Allow the tube to cool in cold running water or by immersing in chilled water. Add 25 ml of diethyl ether (peroxide free), close with bark cork or glass stopper, which is wetted with water before insertion, and shake vigorously for 1 min. Open the tube and then add 25 ml light petroleum ether, close the tube, and shake again vigorously for a minute. Let the tube stand on the flat bottom of the lower bulb until the upper ethereal layer is clear and separated completely from the aqueous layer, usually not less than 30 min, or centrifuge until clear. If there is a tendency to form emulsion, a little alcohol may be added to help separation of the layers.

Decant off as much as possible of the supernatant layer into a suitable flask by gradually bringing the cylindrical bulb of the tube into a horizontal position. When as much as possible has been poured off, wash the outside of the neck of the tube and the cork or stopper with mixed solvent, collecting the rinsing in the flask. With the Mojonnier tube in a vertical position, wash the inside of the neck with 4–5 ml of mixed solvent and decant. Repeat twice extraction of the liquids remaining in the extraction tube using 15 ml of ether and 15 ml of petroleum solvent every time. Distil carefully the solvents from the flask and dry the flask containing the residual milk fat in an air oven at 102 ± 2 °C for 2 h, cool in a desiccator, and weigh. Heat the flask again in the oven for 30 min. Cool in desiccator and weigh. Repeat the process of heating and cooling and weighing until the difference between two successive weights does not exceed 1 mg. Wash out the fat from the flask with petroleum ether carefully leaving any insoluble residue in the flask. Dry the flask in the oven, cool, and weigh as before. The difference in weights represents the weight of fat extracted from the milk.

Make a blank determination using the specified quantities of reagent whenever a sample is used for test. Replace water in place of milk in the blank and deduct the value found from the test sample. If reagent blank is more than 0.5 mg, purify or replace reagents. Difference between duplicate determinations obtained simultaneously by the same analyst should not be more than 0.03 gm fat/100 gm product.

Calculation:

$$\text{Fat (\%w/w)} = \frac{\text{Weight of fat}}{\text{Weight of milk}} \times 100 \tag{1}$$

Titratable Acidity: Milk acidity is generally determined by direct titration.

Reagent: Phenolphthalein indicator solution: Dissolve 1 g of phenolphthalein in 110 ml of ethyl alcohol and add 80 ml water and 0.1 M sodium hydroxide solution until one drop gives faint pink color. Make up to 200 ml with distilled.

Reference Color Solution: Dissolve 1.5 g of cobalt (II) sulfate hydrate in water and dilute to 100 ml.

Method: Pipette 10 ml of milk into each of the two conical flasks, Control (C) and Sample (S). To the control flask, C, add 1 ml reference color solution and stir. Then to sample S, add 1 ml of phenolphthalein indicator solution and titrate with 0.1 M sodium hydroxide, stirring continuously, until the color matches the pink tint of C. Calculate the acidity as lactic acid.

$$1 \text{ ml } 0.1 \text{ M sodium hydroxide} = 0.0090\text{g lactic acid}$$

Proteins Kjeldahl's Methods:

Same as in Chapter 42: Ash value: Ash is determined by incinerating the dried residue of 10 g of milk at a temperature of 525 ± 25 °C. Any black particles that remain can usually be removed by wetting the ash with distilled water, then evaporating off the water on the boiling water bath and reigniting. An ash value 0.73% may be due to presence of added water.

Lactose. For routine purpose lactose is determined volumetrically with chloramine-T. Lactose may also be determined by GLC and HPLC and enzymatically based on ultraviolet detection of NADH formed by enzymatic oxidation.

Citric acid. Citric acid is determined enzymatically. It can also be determined enzymatically by HPLC by reverse phase ion exchange and proprietary column for organic acids.

Viva Voce:

Q1. What is the acidity of freshly drawn milk?

Q2. Which method is used for determining fat in milk?

Q3. What is the importance of butyrometer tube?

Q4. What is the principle of Rose-Gottileb method?

Q5. Temperature required for the determination of ash value.

Q6. Which method is used for determination of citric acid in milk?

Q7. What is the percentage of mineral matter in milk?

Q8. Which method is used for routine analysis of lactose in milk?

Q9. What is titratable acidity?

Q10. Name the reference method used for fat determination in milk.

Chapter 57

Thin Layer Chromatography of Carbohydrates

Chromatography is a physical method available for separation of closely related compound in a mixture. The separation is attained by alterations in the equilibrium distribution of the components between two immiscible phases, i.e., the stationary and the mobile phases. The stationary phase is normally equipped by means of silica, through which the sample mixture permeates under the impact of the mobile phase. There are numeral interactions between the sample and stationary phase, and these have been well exploited to effect the separation of compounds.

Thin-layer chromatographic (TLC) provides qualitative evidence, and with cautious attention to information, it is promising to acquire quantitative data. Thin-layer chromatography is a technique used to separate and identify compounds. A TLC plate may be thin layer of silica stick, glass, or aluminum plate. The silica gel is used as stationary phase, and the solvent or solvent mixture is used as the mobile phase. In the ideal solvent system, the compounds of interest are soluble to different degrees. Separation results from the partition equilibrium of the components in the mixture.

Materials and Reagents: Silica gel G mixed with 0.2 M of sodium acetate or sodium borate, 2–5% w/v solution of carbohydrates (standards and unknown), or Keiselghur G.

Solvent system: 1-butanol/acetone/water or buffer at pH 5 (40:50:10); propanol/ethyl acetate (30:70)∗.

For detecting reagents: Diphenylamine reagent (DPAR): Dissolve 1.5 g diphenylamine in 100 ml of acetic acid and add 2.75 ml of concentrated sulfuric acid. Store the reagent at 2 °C and use within 6 months.

∗For reducing sugars:

(i) Dissolve 0.5 gm of 3,5-dinitrosalicylic acid in 100 ml of 4% w/v sodium hydroxide. Heat and cool before use.

(ii) Prepare an iodine saturated chamber by putting solid iodine crystals in a closed chamber.

Aakanchha Jain et al. *Basic Techniques in Biochemistry, Microbiology and Molecular Biology: Principles and Techniques*, Springer Protocols Handbooks, https://doi.org/10.1007/978-1-4939-9861-6_57,

Procedures:

(1) Take a clean, dust-free TLC chamber and fill it 1/4 with the solvent system.

(2) Saturate the chamber using filter paper detection. Make sure that the lid is closed to avoid delay in saturation.

(3) Use the appropriate stationary phase to make a layer over the plate. Let it dry normally. Activate in oven 10 min before use.

(4) Draw a line nearly 2 cm from the narrow end, using a permanent marker or pencil which remains stable in the solvent.

(5) Apply the spots of samples using different capillaries.

(6) Place the plate in the saturated chamber in a way that the 2 cm line is above the solvent. Air tight the lid using paraffin wax.

(7) Allow the solvent to run upward through capillary action till more than 3/4 plate is wet.

(8) Take out the plate and allow it to dry at room temperature.

(9) Use the detecting reagents to obtain the spots.

(10) Trace the spots immediately in a butter paper.

(11) Mark the center of spot and measure the distance traveled by the sample from the baseline drawn at 2 cm from the bottom.

(12) Also measure the distance traveled by the solvent from the baseline drawn at 2 cm from the bottom.

(13) Calculate the Rf value and compare it with the standard.

(14) Mostly all carbohydrates except aldopentose (purple) and monosaccharide (fructose-red) give brown spot in plate.

Calculations:

$$Rf = \frac{\text{Distance moved by the substance from bottom line}}{\text{Distance moved by the solvent from bottom line}} \quad (1)$$

Standard Rf values some carbohydrates:...................

Applications:

(1) To detect the presence or absence of carbohydrate in unknown sample

(2) To detect the type of carbohydrate in unknown sample

(3) For isolation of carbohydrate from unknown sample

Viva Voce:

Q1. What are carbohydrates?

Q2. Name some monosaccharides.

Q3. Name some disaccharides.

Q4. Name some polysaccharides.

Q5. What are the detecting reagents for carbohydrates in TLC?

Q6. What is meant by TLC?

Q7. Which solvent system is used for TLC of carbohydrates?

Q8. What is the reason to activate the TLC chamber before experiment?

Q9. Give some application of TLC of carbohydrates.

Q10. What is the formula to calculate Rf value?

Chapter 58

Thin Layer Chromatography of Amino Acid

Materials and Reagents: 2% w/v solution of amino acid to be run on stationary phase made of silica gel G.

Solvent system: "SO1" = n-propanol, water = 7:3.

"SO2" = n-butanol, acetic acid, water = 3:1:1.

Detecting reagent: 1.5% w/v ninhydrin in n-butanol mixed with 0.3% v/v acetic acid.

Procedures:

1. Take a clean, dust-free TLC chamber and fill it 1/4 with the solvent system.
2. Saturate the chamber using filter paper detection. Make sure that the lid is closed to avoid delay in saturation.
3. Use the appropriate stationary phase to make a layer over the plate. Let it dry normally. Activate in oven 10 min before use.
4. Draw a line nearly 2 cm from the narrow end, using a permanent marker or pencil which remains stable in the solvent.
5. Apply the spots of samples using different capillaries.
6. Place the plate in the saturated chamber in a way that the 2 cm line is above the solvent. Air tight the lid using paraffin wax.
7. Allow the solvent to run upwards through capillary action till more than 3/4 plate is wet.
8. Take out the plate and use the detecting reagent to obtain the spots.
9. Heat the plate at 110 °C for 10–12 min.
10. Trace the spots immediately in a butter paper.
11. Mark the center of spot and measure the distance traveled by the sample from the baseline drawn at 2 cm from the bottom.

Aakanchha Jain et al. *Basic Techniques in Biochemistry, Microbiology and Molecular Biology: Principles and Techniques*, Springer Protocols Handbooks, https://doi.org/10.1007/978-1-4939-9861-6_58,

12. Also measure the distance traveled by the solvent from the baseline drawn at 2 cm from the bottom.
13. Calculate the Rf value and compare it with the standard.
14. Mostly all amino acids give blue-purple color spot except for some which shows pink.
15. Calculate the Rf value from Eq. (1) in Chap. 57.

Standard Rf values some amino acids:

Name of amino acid	Rf value n-propanol/water = 7:3
Valine	**0.44**
Tyrosine	0.55
Tryptophan	0.61
Threonine	0.30
Serine	0.26.
Proline	0.24
Phenylalanine	0.62
Methionine	0.63
Lysine	0.12
Leucine	0.61
Isoleucine	0.53
Histidine	0.12
Glycine	0.25
Glutamine	0.25
Glutamic acid	0.31
Cysteine	0.37
Aspartic acid	0.16
Asparagine	0.24
Arginine	0.21
Alanine	0.30

Applications:

(1) To detect the presence or absence of amino acid in unknown sample
(2) To detect the type of amino acid in unknown sample
(3) For isolation of amino acid from unknown sample
(4) Use to monitor the progress of reactions

Viva Voce:

Q1. What are amino acids?

Q2. Name some amino acids.

Q3. How many amino acids are there in the body?

Q4. What are essential amino acids?

Q5. What are the detecting reagents for amino acids in TLC?

Q6. What is meant by TLC?

Q7. Which solvent system is used for TLC of amino acids?

Q8. What is the reason to activate the TLC chamber before experiment?

Q9. What is the use of performing TLC of amino acids?

Q10. What is the formula to calculate Rf value?

Chapter 59

Paper Chromatography of Amino Acid

Material and Reagents: 2% *w/v* solution of amino acid to be run on stationary phase, i.e., Whatman filter no. 2.

Solvent system: n-butanol/acetic acid/water = 12:3:5; methanol/pyridine/water = 80:4:20.

Detecting reagent: 1.5% w/v ninhydrin in ethanol mixed with 0.3% v/v acetic acid or 1.5mL pyridine.

Procedures

(1) Take a clean, dust-free glass chamber and fill it 1/4 with the solvent system.

(2) Saturate the chamber using filter paper detection. Make sure that the lid is closed to avoid delay in saturation.

(3) Meanwhile take a fresh Whatman filter paper, and cut it in rectangular shape without touching it barehands. Use gloves.

(4) Draw a line nearly 2 cm from the narrow end, using a permanent marker or pencil which remains stable in the solvent.

(5) Apply the spots of samples using different capillaries.

(6) Place the Whatman filter paper in the saturated chamber in a way that the 2 cm line is above the solvent. Air tight the lid using paraffin wax, as shown in the photograph below.

(7) Allow the solvent to run upwards through capillary action till more than 3/4 paper is wet (Fig. 1).

(8) Take out the paper, use the detecting reagents to obtain the spots, and dry hot oven for few minutes.

(9) Trace the spots immediately using a sharp HB pencil.

(10) Mark the center of spot, and measure the distance traveled by the sample from the baseline drawn at 2 cm from the bottom.

(11) Also mark and measure the distance traveled by the solvent from the baseline drawn at 2 cm from the bottom.

(12) Calculate the Rf value and compare it with the standard.

Aakanchha Jain et al. *Basic Techniques in Biochemistry, Microbiology and Molecular Biology: Principles and Techniques*, Springer Protocols Handbooks, https://doi.org/10.1007/978-1-4939-9861-6_59,

(13) Mostly all amino acids give blue-purple color spot except for some which shows pink.

(14) Calculate the Rf value from Eq. (1) in Chap. 57.

Standard Rf values of some amino acids

Fig. 1 Chamber setup for paper chromatography

Name of amino acid	Rf value in n-butanol/acetic acid/water = 12:3:5
Alanine	0.38
Arginine	0.20
Asparagine	0.5
Aspartic acid	0.24
Cysteine	0.4
Glutamine	0.13
Glutamic acid	0.30
Glycine	0.26
Histidine	0.11
Isoleucine	0.72
Leucine	0.73
Lysine	0.14
Methionine	0.55
Phenylalanine	0.68

(continued)

Name of amino acid	Rf value in n-butanol/acetic acid/water = 12:3:5
Proline	0.43
Serine	0.27
Threonine	0.35
Tryptophan	0.66
Tyrosine	0.45
Valine	0.61

Applications:

(1) To detect the presence or absence of amino acid in unknown sample

(2) To detect the type of amino acid in unknown sample

(3) Use to monitor the progress of reactions

Viva Voce:

Q.1. What are amino acids?

Q.2. Name some amino acids.

Q.3. How many amino acids are there in the body?

Q.4. What are essential amino acids?

Q.5. What are the detecting reagents for amino acids in TLC?

Q.6. What is the principle of paper chromatography?

Q.7. Which solvent system is used for paper chromatography of amino acids?

Q.8. What is the reason to activate the glass chamber before experiment?

Q.9. What is the use of performing paper chromatography of amino acids?

Q.10. What is the formula to calculate Rf value?

Chapter 60

Paper Chromatography of Carbohydrates

Materials and Reagents: Solvent system: 1-butanol/acetone/water or buffer at pH 5 (40:50:10); propanol/ethyl acetate (30:70)*.

Detecting reagents:

(i) Dissolve 0.5 g of 3,5-dinitro salicylic acid in 100 ml of 4% w/v sodium hydroxide. Heat and cool before use.

(ii) Prepare an iodine-saturated chamber by putting solid iodine crystals in a closed chamber.

Procedures:

1. Take a clean, dust-free TLC chamber and fill it 1/4 with the solvent system.
2. Saturate the chamber using filter paper detection. Make sure that the lid is closed to avoid delay in saturation.
3. Meanwhile take a fresh Whatman filter paper, and cut it in rectangular shape without touching it barehands. Use gloves.
4. Draw a line nearly 2 cm from the narrow end, using a permanent marker or pencil which remains stable in the solvent.
5. Apply the spots of samples using different capillaries.
6. Place the plate in the saturated chamber in a way that the 2 cm line is above the solvent. Air tight the lid using paraffin wax.
7. Allow the solvent to run upwards through capillary action till more than 3/4 paper is wet.
8. Take out the plate and allow it to dry at room temperature.
9. Use the detecting reagents to obtain the spots.
10. Trace the spots immediately as they disappear within 30 min of exposure.

Aakanchha Jain et al. *Basic Techniques in Biochemistry, Microbiology and Molecular Biology: Principles and Techniques*, Springer Protocols Handbooks, https://doi.org/10.1007/978-1-4939-9861-6_60,

11. Mark the center of the spot, and measure the distance traveled by the sample from the baseline drawn at 2 cm from the bottom.
12. Also measure the distance traveled by the solvent from the baseline drawn at 2 cm from the bottom.
13. Calculate the Rf value and compare it with the standard.
14. Mostly all carbohydrates give brown spot in paper.
15. Calculate the Rf value from Eq. (1) in Chap. 57.

Applications:

1. To detect the type of carbohydrate in unknown sample
2. For isolation of carbohydrate from unknown sample
3. To detect the presence or absence of carbohydrate in unknown sample

Viva Voce:

Q1. What are carbohydrates?

Q2. Name some monosaccharides.

Q3. Name some disaccharides.

Q4. Name some polysaccharides.

Q5. What are the detecting reagents for carbohydrates in paper chromatography?

Q6. What is meant by paper chromatography?

Q7. Which solvent system is used for paper chromatography of carbohydrates?

Q8. What is the reason to activate the glass chamber before experiment?

Q9. Give some application of paper chromatography of carbohydrates.

Q10. What is the formula to calculate Rf value?

Chapter 61

Fat Characterization

Fats and oils are the principle forms of energy in organisms and are highly reduced complexes and derivatives of fatty acids. Fatty acids are carboxylic acids with hydrocarbon chains of 4–36 carbons and can be saturated or unsaturated. Triglycerides or triacylglycerols are the simplest lipid developed from fatty acids and are composed of three fatty acids each in ester linkage with a single glycerol.

In contrary to lipid, fats and oils are referred as bulk, crude, or refined product extracted from animal products or oilseeds and other plants of lipid content. The term fat indicates extracted lipids that are solid at room temperature, and oil refers to those that are liquid. Nevertheless, the terms, lipid, fat and oil, often are used conversely.

In saturated fat, there is only single bond between carbon atoms in hydrocarbon chain. The fatty acids are saturated with hydrogen; in a saturated fat, as many hydrogen atoms as possible are attached to the carbon skeleton. In unsaturated fat hydrocarbon chain is composed of double bond and has fewer hydrogen molecules. If there is only single double bond in a fatty acid, it is termed as monounsaturated, while if there are multiple double bonds, it is known as polyunsaturated.

The characterization methods for fats, lipids, and oils are categorized into two:

1. Those developed to analyze bulk oils and fats
2. Those concentrating on analysis of foodstuffs and lipid extracts

1 Methods for Characterization of Fats

Numerous methods exist to measure the characteristics of fats and oils. Some methods (e.g., titer test) have limited use for edible oils (in contrast to soaps and industrial oils). Other methods may require special apparatus not commonly available or may have been antiquated by modern instrumental procedures [e.g., volatile acid methods (Reichert-Meissl, Polenske, and Kirschner values) have been replaced

Aakanchha Jain et al. *Basic Techniques in Biochemistry, Microbiology and Molecular Biology: Principles and Techniques*, Springer Protocols Handbooks, https://doi.org/10.1007/978-1-4939-9861-6_61,

largely by determination of fatty acid composition using Gas chromatography.

1.1 Melting Point

Principle: The melting point may be defined in various ways, each corresponding to a different residual amount of solid fat. The capillary tube melting point, also known as the complete melting point or clear point, is the temperature at which fat heated at a given rate becomes completely clear and liquid in a one-end closed capillary. The slip melting point is performed similarly to the capillary tube method and measures the temperature at which a column of fat moves in an open capillary when heated. The dropping point (also called the dropping melting point or Mettler dropping point) is the temperature at which the sample flows through a 0.11 in. hole in a sample cup placed in a specialized furnace. The Wiley melting point measures the temperature at which a $1/8 \times 3/8$ in. disk of fat, suspended in an alcohol-water mixture of similar density, changes into a sphere. A disadvantage of the Wiley melting point is the subjective determination as to when the disk is spherical. A disadvantage of the slip melting point is its 16 h stabilization time.

Applications: The capillary tube method is less useful for oils and fats (in comparison to pure compounds) since they lack a sharp melting point due to their array of various components.

1.2 Iodine Value

Principle: The iodine value (or iodine number) is a measure of degree of unsaturation, which is the number of carbon-carbon double bonds in relation to the amount of fat or oil. Iodine value is defined as the grams of iodine absorbed per 100 g of sample. The higher the amount of unsaturation, the more iodine is absorbed and the higher the iodine value. The calculated iodine value is not meant to be a rapid method but instead gives two results (iodine value of triacylglycerols and free fatty acids [FFAs]) from one analysis (fatty acid composition).

Procedure: A quantity of fat or oil dissolved in solvent is reacted, avoiding light, with a measured amount of iodine or some other halogen such as ICl or IBr. Halogen addition to double bonds takes place. A solution of potassium iodide is added to reduce excess ICl to free iodine. The liberated iodine is then titrated with a sodium thiosulfate standard using a starch indicator, and the iodine value is calculated:

$$\text{Iodine Value} = \frac{(B - S) \times N \times 126.9}{W \times 1000} \times 100 \qquad (1)$$

where:

Iodine value = g iodine absorbed per 100 g of sample

B = volume of titrant (mL) for blank

S = volume of titrant (mL) for sample

N = normality of $Na_2S_2O_3$ (mol/L)

126.9 = MW of iodine (g/mol)

W = sample mass (g)

1000 = conversion of units (mL/L)

Applications: Iodine value is used to characterize oils, to follow the hydrogenation process in refining, and as an indication of lipid oxidation, since there is a decline in unsaturation during oxidation. The calculated value tends to below for materials with a low iodine value and for oils with greater than 0.5 % unsaponifiable material (e.g., fish oils). The Wijs iodine procedure uses ICl, and the Hanus procedure uses IBr. The Wijs procedure may be preferable for highly unsaturated oils as it reacts faster with the double bonds.

1.3 Saponification Value

Principle: Saponification is the process of breaking down or degrading a neutral fat into glycerol and fatty acids by treatment of the fat with alkali. The saponification value (or saponification number) is defined as the amount of alkali necessary to saponify a given quantity of fat or oil. It is expressed as the milligrams of KOH required to saponify 1 g of the sample. The saponification value is an index of the mean molecular weight of the triacylglycerols in the sample. The mean molecular weight of the triacylglycerols may be divided by 3 to give an approximate mean molecular weight for the fatty acids present; the smaller the saponification value, the longer the average fatty acid chain length.

Procedure: Excess alcoholic potassium hydroxide is added to the sample, and the solution is heated to saponify the fat. The unreacted potassium hydroxide is back-titrated with standardized HCl using phenolphthalein as the indicator, and the saponification value is calculated:

$$\text{Saponification value} = \frac{(B - S) \times N \times 56.1}{W} \quad (2)$$

where:

Saponification value = mg KOH per g of sample

B = volume of titrant (mL) for blank

S = volume of titrant (mL) for sample

N = normality of HCl (mmol/mL)

56.1 = MW of KOH (mg/mmol)

W = sample mass (g)

The calculated saponification value is obtained from fatty acid composition using equation given below. The fractional molecular weight of each fatty acid in the sample must be determined first by multiplying the fatty acid percentage (divided by 100) by its molecular weight. The mean molecular weight is the sum of the fractional weights of all the fatty acids in the sample:

$$\text{Calculated saponification value} = \frac{3 \times 56.1 \times 1000}{(\text{Mean molecular weight} \times 3) + 92.09 - (3 \times 8)} \quad (3)$$

where:

Calculated saponification value = mg KOH per g of sample

3 = number of fatty acids per triacylglycerol

56.1 = MW of KOH (g/mol)

1,000 = conversion of units (mg/g)

92.09 = MW of glycerol (g/mol)

18 = MW of water (g/mol)

Saponification value is useful for determining the average fatty acid chain length of an oil or fat. The calculated saponification value is not applicable to fats and oils containing high amounts of unsaponifiable material, FFAs (>0.1 %), or mono- and diacylglycerols (>0.1 %).

Lipid Oxidation: The term rancidity refers to the off odors and flavors resulting from lipolysis (hydrolytic rancidity) or lipid oxidation (oxidative rancidity). Lipid oxidation (also called autoxidation) as it occurs in bulk fats and oils proceeds via a self-sustaining free radical mechanism. Depending on the fatty acid composition, the prooxidants to which it is exposed, the type and amount of antioxidants present, and other factors, the decomposition of fats and oils can proceed via various mechanisms that produce many different compounds.

Sample Preparation: Most methods require lipid extraction prior to analysis. However, variations of some methods may begin with the original foodstuff.

1.4 Peroxide Value

Principle: Peroxide value is defined as the milliequivalents (mEq) of peroxide per kilogram of sample. It is a redox titrimetric determination. The assumption is made that the compounds reacting under the conditions of the test are peroxides or similar products of lipid oxidation.

Procedure: The fat or oil sample is dissolved in glacial acetic acid-isooctane (3:2). Upon addition of excess potassium iodide, which reacts with the peroxides, iodine is produced. The solution then is titrated with standardized sodium thiosulfate using a starch indicator. Peroxide value is calculated as shown below:

$$\text{Peroxide value} = \frac{(S - B) \times N}{W} \times 1000 \quad (4)$$

where:

Peroxide value = mEq peroxide per kg of sample

S = volume of titrant (mL) for sample

B = volume of titrant (mL) for blank

N = normality of $Na_2S_2O_3$ solution (mEq/mL)

1,000 = conversion of units (g/kg)

W = sample mass (g)

Applications: Peroxide value measures a transient product of oxidation. However peroxide values measured on a given sample stored over time increase and then decrease, and products are usually determined (using sensory analysis) to be rancid during the period of increasing values. For determination in foodstuffs, a disadvantage of this method is the 5 g fat or oil sample size required; it is difficult to obtain sufficient quantities from foods low in fat. This method is empirical, and any modifications may change results. Despite its drawbacks, peroxide value is one of the most common tests of lipid oxidation. High-quality, freshly deodorized fats and oils will have a peroxide value of zero. Peroxide values >20 correspond to very poor-quality fats and oils, which normally would have significant off flavors. For soybean oil, peroxide values of 1–5, 5–10, and >10 correspond to low, medium, and high levels of oxidation, respectively.

2 Methods for Lipid Components

The lipid present in food commodities or bulk fats and oils can be characterized by measuring the amount of its various fractions, which include fatty acids; mono-, di-, and triacylglycerols; phospholipids; sterols (including cholesterol and phytosterols); lipid-soluble pigments; and vitamins.

2.1 Fatty Acid Composition and Fatty Acid Methyl Esters (FAMEs)

The fatty acid composition, or fatty acid profile, of a food product is determined by quantifying the kind and amount of fatty acids that are present, usually by extracting the lipids and analyzing them using capillary GC.

Principle: To increase volatility before GC analysis, triacylglycerols are typically transesterified to form fatty acid methyl esters (FAMEs). Acyl lipids are readily transesterified using base such as sodium hydroxide and methanol. Sodium methoxide produced by this combination will create FAMEs from acyl lipids rapidly but will not react with free fatty acids. Acidic reagents such as methanolic HCL or boron trifluoride (BF3) react rapidly with FFAs but more slowly with acyl lipids. Procedures, such as two-step methylation, first react to the lipid with 0.5 *N* NaOH then with excess BF3/methanol. This allows a rapid methylation of FFAs, acyl lipids, and phospholipids. The sodium hydroxide step is not a saponification procedure (i.e., hydrolysis of acyl groups); it is a direct transmethylation.

Procedure: The lipid is extracted from the food, for example, by homogenizing with a suitable solvent such as hexane: isopropanol (3:2, v/v) and then evaporating the solvent. The FAMEs are prepared by combining the extracted lipid with sodium hydroxide methanol and internal standard in isooctane and then heating at 100°C for 5 min. The sample is cooled, and then excess BF3-methanol is added with further heating (100°C for 30 min). After addition of saturated aqueous sodium chloride, additional isooctane, and mixing, the upper isooctane solution containing the FAMEs is removed and dried with anhydrous Na_2SO_4, and then diluted to a concentration of 5–10 % for injection onto the GC. Determination of the fatty acid composition of a product permits the calculation of the following categories of fats that pertain to health issues and food labeling: percent saturated fatty acids, percent unsaturated fatty acids, percent monounsaturated fatty acids, percent polyunsaturated fatty acids, CLAs, and percent *trans*-isomer fatty acids. Calculation of fatty acids as a percentage is referred to as normalization, i.e., the areas of all of the FAMEs are summed, and the percent area of each fatty acid is calculated relative to the total area. This is a reasonable procedure because with flame ionization detectors (FID), the weight of fatty acids in a mixture closely parallels the area on the chromatogram.

2.2 Cholesterol and Phytosterols

Many methods exist for the quantification of cholesterol and phytosterols in various matrices. Consulting research literature will give an indication of current practice and methods that may be less laborious or adapted for use with specific foodstuffs.

Principle: The lipid extracted from the food is saponified. The saponification process is a hydrolysis, with acyl lipids being converted to water-soluble FFA salts. Other components called the unsaponifiable or nonsaponifiable matter do not change in solubility after hydrolysis and thus remain soluble in organic solvents. Cholesterol is extracted and derivatized to form trimethylsilyl (TMS) ethers or acetate esters. This increases their volatility and reduces problems of peak tailing during chromatography. Quantitation is achieved using capillary GC.

Procedure: Lipids are extracted from the food and saponified, and the unsaponifiable fraction is extracted. This is accomplished by filtering an aliquot of the chloroform layer through anhydrous sodium sulfate and evaporating to dryness in a water bath using a stream of nitrogen gas. Concentrated potassium hydroxide and ethanol are added, and the solution is refluxed. Aliquots of benzene and 1 *N* potassium hydroxide are added and then shaken. The aqueous layer is removed, and the process is repeated with 0.5 *N* potassium hydroxide. After several washes with water, the benzene layer is dried with anhydrous sodium sulfate, and an aliquot is evaporated to dryness on a rotary evaporator. The residue is taken up in dimethylformamide. An aliquot of this sample is derivatized by adding examethyldisilazane (HMDS) and trimethylchlorosilane (TMCS). Water (to react with and inactivate excess reagent) and an internal standard in heptane are added, and then the solution is centrifuged. A portion of the heptane layer is injected into a GC equipped with a nonpolar column. The HMDS and TMCS reagents are rapidly inactivated by water, and thus the reaction conditions must remain anhydrous.

Applications: GC quantitation of cholesterol is recommended since many spectrophotometric methods are not specific for cholesterol. In the past, samples such as eggs and shrimp have had their cholesterol contents overestimated by relying on less specific colorimetric procedures. Other GC, HPLC, and enzymatic methods are available. For example, cholesterol methods developed for frozen foods and meat products eliminate the fat extraction step, directly saponifying the sample; compared to the method defined previously, they are more rapid and avoid exposure to toxic solvents. Cholesterol oxidation products as well as phytosterols can be quantified using the GC procedure outlined for cholesterol. A wide range of methods for analysis of sterols exists in the literature; most use TMS ether formation to increase volatility of the hydroxide containing sterols and to improve chromatographic resolution (reduce peak tailing).

Viva Voce:

Q1. Define triglycerides.

Q2. What is the difference in saturated and unsaturated fat?

Q3. Give the name of different methods used for melting point determination of fat.

Q4. What is iodine value?

Q5. What is saponification value?

Q6. What is rancidity?

Q7. Define autoxidation.

Q8. What is peroxide value?

Q9. What is the importance of peroxide value on quality of fat and oil?

Q10. Which method is suitable for quantification of cholesterol and why?

Chapter 62

Analysis of Bread

Yeasts are living unicellular, eukaryotic, polyphyletic, and ubiquitous microorganisms commonly found on fruits, vegetables, and other plant materials. They are facultative anaerobes and can respire and survive under both aerobic and anaerobic conditions. In the absence of oxygen, they can ferment sugar into alcohol (ethanol) and carbon dioxide and low biomass. In well-aerated conditions, the cells could be able to get enough energy and convert sugar into high biomass.

Bread is one of the oldest staple foodstuffs, which is made and eaten in most countries around the world. It may be described as a fermented confectionary product produced mainly from wheat flour, water, yeast, and salt by a series of process involving mixing, kneading, proofing, shaping, and baking. Bread making is fundamentally a temperature dependent of two-step progression, consisting of fermentation, in which CO_2 production linked with yeast activity is manifested in porous dough structure with the development of dough volume during baking where yeast activity is ended and the bread structure is finalized. The concluding bread structure depends on dough ingredients, yeast activity, fermentation temperature, and gas bubble formation. Many types of breads are baked with baker's yeast.

1. Yeast is mixed with warm water, flour, and salt. The dough so formed is thoroughly kneaded, and the mass kept warm for some hours.
2. By this time the yeast cells convert some of the starch into glucose and into ethanol and CO_2. The enzyme of the flour also converts some of the starch into maltose and some glucose. More flour is added, and the process is allowed to continue for some hour longer, when the formation of CO_2 causes the mass to rise.
3. The dough is heated to about 230 °C for 90 min. The heat expands the gas bubble killing the yeast and causes the outer part of the dough to be hardened by coagulating the protein. It

Aakanchha Jain et al. *Basic Techniques in Biochemistry, Microbiology and Molecular Biology: Principles and Techniques*, Springer Protocols Handbooks, https://doi.org/10.1007/978-1-4939-9861-6_62,

also converts starch into soluble starch and dextrins and forming the crust. The brown appearance of this is due to conversion of glucose into caramel by the action of heat.

Detection of Starch and Dextrin:

1. A piece of crumb of a stale white loaf is taken, rubbed up finely, and pounded with cold water in a mortar.
2. It is strained and squeezed through muslin.
3. A white fluid containing wheat starch grains is obtained. The fluid is filtered.
4. A little Fehling's solution is added to a portion of the filtrate and boiled; a well-marked reduction occurs. A drop or two of iodine is added to another portion; a purple color is produced showing the presence of dextrins. If two drops of very dilute iodine is added, a pure blue color is produced showing that a small amount of soluble starch is also present.
5. The above procedure is repeated using the crust instead of crumb. It is noted that reducing sugar is absent or present in traces only; dextrins and starch are present.

Viva Voce:

Q1. What are some differences between quick breads and yeast breads?

Q2. What nutrients do you get from the grains in our breads?

Q3. What is the role of yeast when breads are baked?

Q4. Which reagent is used for detection of dextrin in bread?

Q5. How to detect the presence of soluble starch present in bread?

Q6. Which color is visible in presence of starch?

Q7. Which color is visible in presence of dextrin when Fehling reagent is added?

Q8. What is fermentation?

Q9. Why the outer part of bread get hardened?

Q10. How will you determine presence or absence of reducing sugar in bread?

Appendices

Appendix I: Composition of Different Types of Media

(I) Simple Media

Peptone broth

- Peptone: 20 g
- Distilled water: 500 ml

Nutrient broth

- Beef extract: 1.5 g
- Peptone: 2.5 g
- Distilled water: 500 ml
- 10 N NaOH for pH adjustment

Nutrient agar

- Beef extract: 1.5 g
- Peptone: 2.5 g
- Agar: 7.5 g
- Distilled water: 500 ml
- 10 N NaOH for pH adjustment

Lactose agar

- Beef extract: 1.5 g
- Peptone: 2.5 g
- Lactose: 2.5 g
- Distilled water: 500 ml

Soft nutrient agar

- Beef extract: 1.5 g
- Peptone: 2.5 g
- Agar: 3.5 g
- Distilled water: 500 ml
- 10 N NaOH for pH adjustment

Aakanchha Jain et al. *Basic Techniques in Biochemistry, Microbiology and Molecular Biology: Principles and Techniques*,
Springer Protocols Handbooks, https://doi.org/10.1007/978-1-4939-9861-6,

Nutrient gelatin
- Beef extract: 1.5 g
- Peptone: 2.5 g
- Gelatin: 20 g
- Distilled water: 500 ml

(II) Differential Media

Eosin methylene blue agar (EMB agar) [for growth of gram-negative bacteria/for lactose and sucrose fermentation]
- Peptone: 5gm
- Sucrose: 2.5 g
- Lactose: 2.5 g
- Dipotassium hydrogen phosphate: 1 g
- Eosin Y: 0.2 g
- Methylene blue: 0.03gm
- Agar: 7.5 g
- Distilled water: 500 ml

Starch agar
- Peptone: 5 g
- Yeast extract: 5 g
- Dipotassium hydrogen phosphate: 2.5 g
- Soluble starch: 1.5 g
- Agar: 7.5 g
- Distilled water: 500 ml

Blood agar
- Beef extract: 1.5 g
- Peptone: 2.5 g
- Sodium chloride: 1.5 g
- Agar: 7.5 g
- Distilled water: 500 ml
- pH adjusted 7.3

Mannitol salt agar (MSA) [for gram-positive bacteria, isolation of pathogenic staphylococci]
- Enzymatic digest of casein
- Enzymatic digest of animal tissue
- Beef extract (source of nitrogen, vitamin, and carbon)
- D-Mannitol
- Sodium chloride: 7.5%
- Phenol red (indicator)
- Agar
- Final pH: 7.4

(III) Selective Media

Nutrient agar with crystal violet
- Beef extract: 1.5 g
- Peptone: 2.5 g

Agar: 7.5 g
Crystal violet 0.1%
Distilled water: 500 ml
10 N NaOH for pH adjustment

(IV) Differential-Selective Media

MacConkey agar [for growth of gram-negative bacteria]

Peptone: 8.5 g
Lactose: 5 g
Bile salts: 0.75 g
Protease peptone: 1.5 g
Sodium chloride: 2.5 g
Neutral red: 0.015 g
Crystal violet: 0.001 g
Agar: 6.75 g
Distilled water: 500 ml
Adjusted to pH 7.1

Appendix II: Buffers

(1) Preparation of Citrate Buffer (pH 3.0–6.2)

Solution A: 0.1M citric acid monohydrate (FW = 210.14)
Solution B: 0.1M trisodium citrate, dihydrate (FW = 294.12)

pH	Amount of solution A (ml)	Amount of solution B (ml)
3.0	82.0	18.0
3.2	77.5	22.5
3.4	73.0	27.0
3.6	68.5	31.5
3.8	63.5	36.5
4.0	59.0	41.0
4.2	54.0	46.0
4.4	49.5	50.5
4.6	44.5	55.5
4.8	40.0	60.0
5.0	35.0	65.0

(continued)

pH	Amount of solution A (ml)	Amount of solution B (ml)
5.2	30.5	69.5
5.4	25.5	74.5
5.6	21.0	79.0
5.8	16.0	84
6.0	11.5	88.5
6.2	8.0	92.0

(2) Preparation of Phosphate Buffer (pH 5.8–8.0)

Prepare 100ml of a 0.1M phosphate buffer by mixing sodium phosphate, dibasic dihydrate, and sodium phosphate monobasic monohydrate, and dilute to 100ml with distilled water.

Solution A: 0.2M sodium phosphate, dibasic dihydrate (FW = 178.05)

Solution B: 0.2M sodium phosphate, monobasic, monohydrate (FW = 138.01)

pH	Amount of solution A (ml)	Amount of solution B (ml)
5.8	4.0	46.0
6.0	6.15	43.85
6.2	9.25	40.75
6.4	13.25	36.75
6.6	18.75	31.25
6.8	24.25	25.5
7.0	30.5	19.5
7.2	36.0	14.0
7.4	40.5	9.5
7.6	43.5	6.5
7.8	45.75	4.25
8.0	47.35	2.65

(3) Preparation of Tris-HCl buffer

Prepare Tris-HCl buffer by mixing equimolar solution of 0.2M Tris and 0.2M HCl. Mix both the solution in proportion as described below and dilute to 200ml.

Solution A: 0.2M Tris (MW121.14)

Solution B: 0.2M HCl

pH	Amount of solution A (ml)	Amount of solution B (ml)
7.2	50	44.2
7.4	50	41.4
7.6	50	38.4
7.8	50	32.5
8.0	50	26.8
8.2	50	21.9
8.4	50	16.5
8.6	50	12.2
8.8	50	8.1
9.0	50	5.0

(4) Preparation of Bicarbonate-Carbonate Buffer (pH 9.2–10.8)

To prepare 100ml of a 0.1M bicarbonate buffer solution, mix sodium bicarbonate and sodium carbonate, decahydrate, as given below.

Solution A: 0.1M sodium bicarbonate (MW = 84.0)

Solution B: 0.1M sodium carbonate, decahydrate (FW = 286.2)

pH	Amount of solution A (ml)	Amount of solution B (ml)
9.1	80	20
9.4	70	30
9.5	60	40
9.7	50	50
9.9	40	60
10.1	30	70
10.3	20	80
10.6	10	90

(5) Preparation of Sodium Acetate-Acetic Acid Buffer (pH 3.6–5.8)

To prepare 100ml of a 0.1M acetate buffer, mix equimolar solution of sodium acetate and acetic acid in proportions mentioned below.

Solution A: 0.1M of sodium acetate (MW= 82.04)

Solution B: 0.1M acetic acid

pH	Amount of solution A (ml)	Amount of solution B (ml)
3.6	7.5	92.5
3.8	12.0	88.0
4.0	18.0	82.0
4.2	26.5	73.5
4.4	37.0	63.0
4.6	48.0	52.0
4.8	59.0	41.0
5.0	70.0	30.0
5.2	79.0	21.0
5.4	86.0	14.0
5.6	91.0	0.9
5.8	94.0	0.6

(6) Preparation of 1M HEPES Buffer (pH 7.5)

Mix 23.83 g HEPES in 100ml water; to adjust pH to 7.5, add 0.1 M sodium hydroxide solution dropwise, and mix thoroughly. Measure pH after addition of every 1 ml of KOH.

(7) Preparation of Phosphate Buffer Saline (PBS pH 7.2–7.4)

NaCl – 8.0 g

Na_2HPO_4 – 1.44 g

K_2HPO_4 – 0.24 g

Mix all the salts in 800 ml distilled water, adjust pH to 7.2/ 7.4 with the help of 1N HCl with shaking, and make up the volume to 1000 ml with distilled water.

(8) Preparation of Phosphate Buffer Saline Tween 20 (PBST, pH 7.4)

Mix 137 mM NaCl, 0.7 mM KCl, 8.1 mM Na_2HPO_4, 1.47 mM K_2HPO_4, and 0.05% Tween 20. Adjust pH to 7.4 with the help of 1N HCl.

(9) Preparation of 50X TAE

Mix 242 g of Tris base, 37.2 g sodium EDTA dehydrated, 800 ml distilled water, and 57.1 ml acetic acid, adjust pH, make up the volume to 1000 ml, filter, sterilize, and store at 4°C.

(10) Preparation of Tris EDTA (TE pH 8.0)

Mix 1mM EDTA and 10mM Tris-HCl; adjust pH with 1 N HCl.

(11) Preparation of 10X Tris Borate EDTA Buffer (pH 8.0)

Mix 108 g of Tris base, 55 g of boric acid, and 40 ml 0.5M EDTA in 800 ml of double distilled water, adjust pH with 1 N HCl, make up the volume to 1000ml, filter, sterilize, and store at 4°C.

(12) Preparation of Tris EDTA Sodium Chloride Buffer (TEN, pH 8.0)

Mix 40 mM Tris-HCl (pH 7.5), 1mM EDTA (pH 8.0), and 150 mM NaCl; adjust pH with 1N NaOH.

Abbreviation: MW = molecular weight, FW = formula weight

Bibliography

1. Patel AH (1984) Industrial microbiology, 1st edn. Macmillan India Limited, New Delhi, pp 96–112
2. Bartoli E, Fra GP, Schianca GC (2011) The oral glucose tolerance test (OGTT) revisited. Eur J Intern Med 22(1):8–12
3. British Pharmacopoeia (2011) The Stationery office British Pharmacopoeia
4. Chang SK, Zhang Y (2017) Protein analysis. In: Food analysis. Springer, Cham, pp 315–331
5. Chawla R. 2014 Practical clinical biochemistry: methods and interpretations. New Delhi JP Medical Ltd; 42, 43, 217-220,
6. Cui G, Florholmen J (2008) Polarization of cytokine profile from Th1 into Th2 along colorectal adenoma-carcinoma sequence: implications for the biotherapeutic target. Inflamm Allergy Drug Targets 7:94–97
7. Dandekar SP, Rane SA (2004) Practicals and viva in medical biochemistry. Elsevier, New Delhi
8. David T Plummer (1988) An introduction to practical biochemistry, 3rd edn. TATA McGraw Hill Publishing Company Ltd., New Delhi, vol 28, pp 57-64 reprint, 1988.
9. Deb AC (2013) Comprehensible viva and practical biochemistry. New Central Book Agency, Kolkata, pp 5–10
10. Evans JR (1905) A laboratory handbook for the analysis of milk, butter and cheese
11. Fischbach F (2004) A manual of laboratory and diagnostic tests, 7th edn. Lippincott Williams & Wilkins, Philadelphia
12. https://en.wikipedia.org/wiki/Interleukin. Date access: 7 May 2019
13. Indian Pharmacopoeia (2010) Govt. of India, ministry of health and family welfare. Indian Pharmacopoeial Commission, Ghaziabad, Delhi
14. Indian Pharmacopoeia (2010) Govt. of India, The Controller Of Publication, Delhi, India
15. Isenberg HD Clinical microbiology procedures handbook, 2nd edn
16. Owen JA, Iggo B, Scandrett FJ, Stewart CP (1954) The determination of creatinine in plasma or serum, and in urine; a critical examination. Biochem J 58:426–437
17. Kale MS, Kale MR (2008) Practical biochemistry & clinical pathology. CBS, Delhi
18. Kantola T, Klintrup K, Vayrynen JP, Vornanen J, Bloigu R, Karhu T, Herzig KH, Napankangas J, Makela J, Karttunen TJ, Tuomisto A, Makinen MJ (2012) Stage-dependent alterations of the serum cytokine pattern in colorectal carcinoma. British J Cancer 107:1729–1736
19. Welte K, Wang CY, Mertelsmann R, Venuta S, Feldman SP, Moore MAS (1982) Purification of human interleukin 2 to apparent homogeneity and its molecular heterogeneity. J Exp Med 156:454–464
20. Lapage S, Shelton J, Mitchell T (1970) In: Norris J, Ribbons D (eds) Methods in microbiology, vol 3A. Academic Press, London
21. Malhotra VK (2003) Practical biochemistry for students. Jaypee Bros. Medical Publications, New Delhi, pp 9–19

22. Morgan DA, Ruscetti FW, Gallo RC (1976) Selective in vitro growth of Tlymphocytes from normal human bone marrows. Science (Wash D C) 193:1007
23. Mu P, Plummer DT (2001) Introduction to practical biochemistry. Tata McGraw-Hill Education, New Delhi
24. N.I.H. Memorandum (1955) Culture media for sterility tests, 4th Revision
25. Gunasekaran P (2000) Laboratory manual in microbiology. New Age International Publishers. Reprint 2000, pp 91–97.
26. Pike OA, O'Keefe S (2017) Fat characterization. InFood analysis. Springer, Cham, pp 407–429
27. Godkar PB, Godkar DP (2014) Textbook of medical laboratory technology: clinical laboratory science and molecular diagnostics, vol 2, 3rd edn. Bhalani Publishing House, Mumbai
28. Gaud RS, Gupta GD, Gokhale SB (2010) In: Prakshan N (ed) Practical biotechnology, 6th edn
29. Vyas SP, Kohli DV (2009) Pharmaceutical biochemistry, 1st edn. CBS Publishers and Distributors, New Delhi
30. Owen JA, Iggo B, Scandrett FJ, Stewart CP (1954) The determination of creatinine in plasma or serum, and in urine; a critical examination. Biochem J 58(3):426–437
31. The United States Pharmacopoeia (2014) The United States Pharmacopoeial Convention. Rockville.
32. USP 35 NF 30 (2010) United States Pharmacopoeia. The United States Pharmacopoeial Convention, Rockville
33. Varley H (1954) Practical clinical biochemistry. Heinemann Medical, London
34. Kale V, Bhusari K (2005) Practical microbiology: principles and techniques, vol 63. Himalaya Publishing House, Bombay, pp 51–52. Revised First Edition
35. Vyas SP, Kohli DV (2009) Pharmaceutical biochemistry. CBS Publishers & Distributors, New Delhi, pp 209–228
36. WHO Diabetes criteria (1999)

MIX
Papier aus verantwortungsvollen Quellen
Paper from responsible sources
FSC® C105338

If you have any concerns about our products,
you can contact us on
ProductSafety@springernature.com

In case Publisher is established outside the EU,
the EU authorized representative is:
Springer Nature Customer Service Center GmbH
Europaplatz 3, 69115 Heidelberg, Germany

Printed by Libri Plureos GmbH
in Hamburg, Germany